Glencoe McGraw-Hill

Math Triumphs

Book 2: Number and Operations

Authors

Basich Whitney • Brown • Dawson • Gonsalves • Silbey • Vielhaber

McGraw Hill Glencoe

Photo Credits

All coins photographed by United States Mint.
All bills photographed by Michael Houghton/StudiOhio.
Cover Peter Sterling/Getty Images; **iv** (1 7 8)File Photo, (2 3)The McGraw-Hill Companies, (4 5 6)Doug Martin; **vi** Getty Images; **vii** PunchStock; **viii** CORBIS; **186-187** John Carnemolla/CORBIS; **193** CORBIS; **194** Getty Images; **198** Superstock; **199** Getty Images; **201** PunchStock; **206** CORBIS; **208** Alamy; **214** CORBIS; **220** Alamy; **221, 229** Getty Images; **232–233** Koji Aoki/Jupiterimages; **240** (t)Greg Fiume/CORBIS, (b)Jupiterimages; **245, 248** Getty Images; **261** Fotosearch; **262** Jules Frazier/Getty Images; **266** CORBIS; **272** Ryan McVay/Getty Images; **281** Don Hammond/CORBIS; **284–285** CORBIS; **285** (tl)Arthur Morris/CORBIS, (tr)Adam Jones/Getty Images, (b)Mark Ransom; **289** PunchStock; **292** Lars A. Niki/The McGraw-Hill Companies; **297** Getty Images; **304** (t)Millard H. Sharp/Photo Researchers, Inc., (b)Steve Maslowski/Visuals Unlimited; **306** G.K. & Vikki Hart/Getty Images; **313** Digital Vision.

The McGraw·Hill Companies

Macmillan/McGraw-Hill
Glencoe

Send all inquiries to:
Glencoe/McGraw-Hill
8787 Orion Place
Columbus, OH 43240-4027

ISBN: 978-0-07-888208-1
MHID: 0-07-888208-7

Printed in the United States of America.

1 2 3 4 5 6 7 8 9 10   066   17 16 15 14 13 12 11 10 09 08

Math Triumphs
Grade 6, Book 2

Math Triumphs

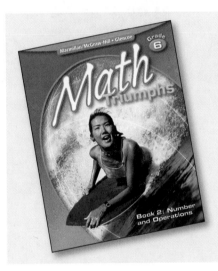

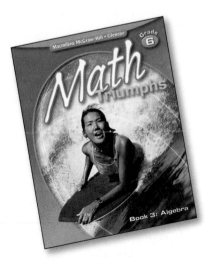

Authors and Consultants

AUTHORS

Frances Basich Whitney
Project Director, Mathematics K–12
Santa Cruz County Office of Education
Capitola, California

Kathleen M. Brown
Math Curriculum Staff Developer
Washington Middle School
Long Beach, California

Dixie Dawson
Math Curriculum Leader
Long Beach Unified
Long Beach, California

Philip Gonsalves
Mathematics Coordinator
Alameda County Office of Education
Hayward, California

Robyn Silbey
Math Specialist
Montgomery County Public Schools
Gaithersburg, Maryland

Kathy Vielhaber
Mathematics Consultant
St. Louis, Missouri

CONTRIBUTING AUTHORS

Viken Hovsepian
Professor of Mathematics
Rio Hondo College
Whittier, California

FOLDABLES Study Organizer **Dinah Zike**
Educational Consultant,
Dinah-Might Activities, Inc.
San Antonio, Texas

CONSULTANTS

Assessment

Donna M. Kopenski, Ed.D.
Math Coordinator K–5
City Heights Educational Collaborative
San Diego, California

Instructional Planning and Support

Beatrice Luchin
Mathematics Consultant
League City, Texas

ELL Support and Vocabulary

ReLeah Cossett Lent
Author/Educational Consultant
Alford, Florida

Reviewers

Each person below reviewed at least two chapters of the Student Edition, providing feedback and suggestions for improving the effectiveness of the mathematics instruction.

Patricia Allanson
Mathematics Teacher
Deltona Middle School
Deltona, Florida

Debra Allred
Sixth Grade Math Teacher
Wiley Middle School
Leander, Texas

April Chauvette
Secondary Mathematics Facilitator
Leander Independent School District
Leander, Texas

Amy L. Chazarreta
Math Teacher
Wayside Middle School
Fort Worth, Texas

Jeff Denney
Seventh Grade Math Teacher, Mathematics
    Department Chair
Oak Mountain Middle School
Birmingham, Alabama

Franco A. DiPasqua
Director of K-12 Mathematics
West Seneca Central
West Seneca, New York

David E. Ewing
Teacher
Bellview Middle School
Pensacola, Florida

Mark J. Forzley
Eighth Grade Math Teacher
Westmont Junior High School
Westmont, Illinois

Virginia Granstrand Harrell
Education Consultant
Tampa, Florida

Russ Lush
Sixth Grade Math Teacher, Mathematics
    Department Chair
New Augusta - North
Indianapolis, Indiana

Joyce B. McClain
Middle School Math Consultant
Hillsborough County Schools
Tampa, Florida

Suzanne D. Obuchowski
Math Teacher
Proctor School
Topsfield, Massachusetts

Karen L. Reed
Sixth Grade Pre-AP Math
Keller ISD
Keller, Texas

Deborah Todd
Sixth Grade Math Teacher
Francis Bradley Middle School
Huntersville, North Carolina

Susan S. Wesson
Teacher (retired)
Pilot Butte Middle School
Bend, Oregon

Contents

Chapter 5 — Multiplication

Glacier Bay National Park, Alaska

Contents

Chapter 6 — Division

Oak Alley Plantation, Louisiana

Contents

Chapter 7

Ratios, Rates, and Unit Rates

Iowa State Capitol, Iowa

SCAVENGER HUNT

Let's Get Started

Use the Scavenger Hunt below to learn where things are located in each chapter.

1. What is the title of Lesson 5-2?

2. What is the Key Concept of Lesson 6-4?

3. On what page can you find the vocabulary term equivalent ratios in Lesson 7-2?

4. What are the vocabulary words for Lesson 5-4?

5. How many Examples are presented in the Chapter 5 Study Guide?

6. What strategy is used in the Step-by-step Problem Solving Practice box on page 290?

7. Describe the art for Exercise #2 on page 299.

8. What headings are used in the table in the Step-by-Step Practice on page 303?

9. On what pages will you find the Study Guide for Chapter 6?

10. In Chapter 7, find the logo and Internet address that tells you where you can take the Online Readiness Quiz.

Chapter 5

Multiplication

How do you use multiplication?

Think about how many miles you travel a day around the city. Think about how many times you travel in a week, a month, or a year. You can use multiplication to find the answer.

STEP 1 Quiz

Math Online ▷ Are you ready for Chapter 5? Take the Online Readiness Quiz at *glencoe.com* to find out.

STEP 2 Preview

Get ready for Chapter 5. Review these skills and compare them with what you will learn in this chapter.

What You Know	What You Will Learn
You know how to add.	*Lessons 5-1, 5-2, and 5-3*

Examples: $5 + 5 + 5 + 5 = 20$
$2 + 2 + 2 = 6$

Lessons 5-1, 5-2, and 5-3

Multiplication is repeated addition.

$5 + 5 + 5 + 5 = 5 \times 4$

You can use **arrays** to model multiplication.

TRY IT!

1. $3 + 3 =$ _____

2. $10 + 10 + 10 + 10 =$ _____

3. $6 + 6 + 6 =$ _____

4. $4 + 4 + 4 + 4 =$ _____

$7 \times 11 =$

You know how to skip count.

Lesson 5-3

Example:
Skip count by 5s.
0, 5, 10, 15, 20, 25, 30, 35, 40, 45, 50,…

Multiples of 8 are the numbers you say when you skip count by 8s.
0, 8, 16, 24, 32, 40, 48, 56, 64, 72…

The multiples of 8 are the multiplication facts below.

TRY IT!

5. Skip count by 4s.

_____

6. Skip count by 6s.

_____

$0 \times 8 = 0$	$5 \times 8 = 40$
$1 \times 8 = 8$	$6 \times 8 = 48$
$2 \times 8 = 16$	$7 \times 8 = 56$
$3 \times 8 = 24$	$8 \times 8 = 64$
$4 \times 8 = 32$	$9 \times 8 = 72$
	$10 \times 8 = 80$

Multiply by 0, 1, 5, and 10

KEY Concept

The **Zero Property of Multiplication** states that any number multiplied by zero is zero.

$$12 \times 0 = 0$$

The **Identity Property of Multiplication** states that any number multiplied by 1 is equal to that number.

$$12 \times 1 = 12$$

You can skip count by 5s to multiply by 5.

$$\overset{1}{5}, \overset{2}{10}, \overset{3}{15}, \overset{4}{20}, \overset{5}{25}, \overset{6}{30}, \overset{7}{35}, ...$$

$$5 \times 4 = 20$$

Multiplying by 10 is similar to multiplying by 1.

$$15 \times 1 = 15$$

$$15 \times 10 = 150$$

When multiplying by 10, place a zero in the ones place.

VOCABULARY

factor
a number that divides into a whole number evenly; also a number that is multiplied by another number

Identity Property of Multiplication
property that states that the product of a factor and 1 equals the factor

multiplication
an operation on two numbers to find their product; it can be thought of as repeated addition

product
the answer to a multiplication problem

Zero Property of Multiplication
property that states any number multiplied by zero is zero

Example 1

Find the product of 38 and 1.

1. What are the factors?
   38 and 1

2. What property can be used to find the product?
   Identity Property of Multiplication

   Any number multiplied by **1** is that number.

3. Write the product.
   38 × 1 = 38

YOUR TURN!

Find the product of 512 and 0.

1. What are the factors?

   _____

2. The property that can be used to find the product is the

   _____.

   Any number multiplied by _____ is _____.

3. Write the product.

   _____

Example 2

Find the product of 29 and 10.

1. Rewrite the problem.

   29 × 10

2. Drop the zero from the factor 10. Then use the Identity Property of Muliplication.

   29 × 1 = 29

3. Add a zero to the product of 29 × 1.

   290

4. Write the product.

   29 × 10 = 290

YOUR TURN!

Find the product of 47 and 10.

1. Rewrite the problem.

   _____

2. Drop the zero from the factor 10. Then use the Identity Property of Muliplication.

   _____

3. Add a zero to the product of 47 × 1.

   _____

4. Write the product.

   47 × 10 = _____

GO ON

Example 3

Find the product of 5 and 14.

1. Rewrite the problem in a vertical format.

2. Multiply the number in the ones column by 5. $5 \times 4 = 20$
   Write the tens digit above the tens column.
   Write the ones digit under the ones column as part of the product.

$$\begin{array}{r} {}^{2} \\ 14 \\ \times\,5 \\ \hline 0 \end{array}$$

3. Multiply 5 times the digit in the tens column. $5 \times 1 = 5$
   Add the 2 regrouped tens for a total of **7** tens.
   The product is 70.

$$\begin{array}{r} {}^{2} \\ 14 \\ \times\,5 \\ \hline 70 \end{array}$$

4. Skip count to check. 14×5 is the 14th multiple of 5.

1	2	3	4	5	6	7	8	9	10	11	12	13	14
5	10	15	20	25	30	35	40	45	50	55	60	65	70

   The 14th multiple of 5 is 70, so the answer makes sense.

YOUR TURN!

Find the product of 5 and 12.

1. Rewrite the problem in a vertical format.

2. Multiply the number in the ones column by 5. $5 \times 2 =$ _____
   Write the tens digit above the tens column.
   Write the ones digit under the ones column as part of the product.

$$\begin{array}{r} 12 \\ \times\,5 \\ \hline \end{array}$$

3. Multiply 5 times the digit in the tens column. $5 \times 1 =$ _____

   Add in the 1 regrouped ten for a total of _____ tens.

   The product is _____.

$$\begin{array}{r} {}^{1} \\ 12 \\ \times\,5 \\ \hline 0 \end{array}$$

4. Skip count to check. 12×5 is the _____ multiple of 5.

   5, 10, 15, 20, 25, _____, _____, _____, _____, _____, _____, _____

   The 12th multiple of 5 is _____, so the answer makes sense.

Who is Correct?

Find the product of 10 and 18.

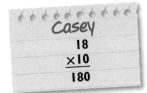
Casey
18
×10
180

Maya
18
×10
108

Jerome
18
×10
1,800

Circle correct answer(s). Cross out incorrect answer(s).

 Guided Practice

Use the Zero Property or the Identity Property of Multiplication to find each product.

1 Find the product of 115 × 0.

Which property should you use?

_____

Write the product. _____

2 Find the product of 59 × 1.

Which property should you use?

_____

Write the product. _____

3 86 × 10 = ⬚⬚0

4 246 × 10 = 2,46⬚

Step by Step Practice

5 Find the product of 5 and 42.

Step 1 Rewrite the problem in a vertical format.

Step 2 Multiply the number in the ones column by 5.

$5 \times 2 =$ _____
Write the tens digit above the tens column.
Write the ones digit under the ones column as part of the product.

42
× 5
___

Step 3 Multiply 5 times the digit in the tens column.

$5 \times 4 =$ _____

Add the 1 regrouped ten for a total of _____ tens.

42
× 5
___

Step 4 Write the product. $5 \times 42 =$ _____

GO ON

Find each product. Show your work.

6 $\begin{array}{r} 93 \\ \times\ 5 \\ \hline \end{array}$

7 $\begin{array}{r} 74 \\ \times\ 5 \\ \hline \end{array}$

8 $\begin{array}{r} 17 \\ \times\ 5 \\ \hline \end{array}$

9 $\begin{array}{r} 34 \\ \times\ 5 \\ \hline \end{array}$

Step by Step Problem-Solving Practice

Solve.

10 **GAMES** Chen and his family are playing a new board game. The directions say for each spin that lands on red, move 5 spaces. Chen has landed on red 12 times during the game. How many total spaces has he moved when he landed on red?

Problem-Solving Strategies
☐ Draw a model.
☐ Use logical reasoning.
☑ Make a table.
☐ Solve a simpler problem.
☐ Work backward.

Understand   Read the problem. Write what you know.

For each spin that lands on red, the

player moves _____ spaces.

Chen landed on red _____ times
during the game.

Plan   Pick a strategy. One strategy is to make a table.

Solve   Use a table to skip count by 5s to find the product. In the first row, write the numbers 1 to 12 in each column. In the second row, skip count by 5s until you get to the number 12.

1	2	3	4								
5	10	15	20								

Write the product.

$12 \times 5 =$ _____

Check   Use vertical multiplication to check.

Copyright © Glencoe/McGraw-Hill, a division of The McGraw-Hill Companies, Inc.

11 SCHOOL Tyson is passing out one pencil to each student before a test. There are 5 rows of students. Each row has 9 students. How many pencils will be passed out? Check off each step.

_____ Understand: I underlined key words.

_____ Plan: To solve the problem, I will

_____.

_____ Solve: The answer is _____.

_____ Check: I checked my answer by _____.

SCHOOL Tyson is passing out pencils for a test.

12 PETS Kanita's pet hamster eats 5 food pellets each day. How many food pellets will Kanita's hamster eat in 30 days?

_____

13 Reflect How is the product of a number multiplied by 10 similar to the number itself?

_____

_____

▶ Skills, Concepts, and Problem Solving

Find each product.

14 $8 \times 1 =$ _____

15 $6 \times 0 =$ _____

16 $5 \times 6 =$ _____

17 $10 \times 42 =$ _____

18 $0 \times 36 =$ _____

19 $1 \times 50 =$ _____

20 $52 \times 5 =$ _____

21 $5 \times 97 =$ _____

22 $10 \times 81 =$ _____

23 $13 \times 1 =$ _____

24 $178 \times 0 =$ _____

25 $5 \times 10 =$ _____

GO ON

Find each product. Show your work.

26 14
    × 5

27 73
    × 10

28 573
    × 1

29 206
    × 10

30 190
    × 5

31 841
    × 10

Solve.

32 PIANO   Susana practices the piano for 45 minutes each day. How many minutes will Susana practice in 10 days?

_____

33 SCHOOL SUPPLIES   Edwin is buying school supplies. He buys 5 packs of paper. There are 250 pages of paper in each pack. How many pages of paper did Edwin buy?

_____

Vocabulary Check   **Write the vocabulary word that completes each sentence.**

34 _____ can be thought of as repeated addition.

35 The answer to a multiplication problem is the _____.

36 Writing in Math   When might you need to multiply a number by zero? Give an example and explain.

_____

_____

_____

STOP

Multiply by 2, 3, 4, and 6

KEY Concept

Use multiples and skip counting when multiplying by 2, 3, 4, and 6.

Multiples of 2, 3, 4, and 6									
×	1	2	3	4	5	6	7	8	9	10
2	2	4	6	8	10	12	14	16	18	20
3	3	6	9	12	15	18	21	24	27	30
4	4	8	12	16	20	24	28	32	36	40
6	6	12	18	24	30	36	42	48	54	60

You should practice memorizing the multiplication facts of 2, 3, 4, and 6.

VOCABULARY

array
objects or symbols displayed in rows of the same length and columns of the same length; the length of a row might be different from the length of a column

factor
a number that divides into a whole number evenly; also a number that is multiplied by another number

multiple
the product of the number and any whole number

product
the answer to a multiplication problem

Example 1

Draw an array to model the expression 2 × 9. Find the product.

1. The first factor is 2, so there will be 2 rows.
   The second factor is 9, so there will be 9 columns, or 9 in each row.

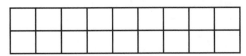

2 × 9

2. Label the array 2 × 9. Count the rectangles.   18

3. Write the multiplication fact.
   2 × 9 = 18

GO ON

YOUR TURN!

Draw an array to model the expression 3 × 8. Find the product.

1. The first factor is _____, so there will be _____ rows.

   The second factor is _____, so there will be _____ columns.

2. Label the array 3 × 8. There are _____ rectangles.

3. Write the multiplication fact. _____ × _____ = _____

Find the product of 3 and 24.

1. Rewrite the problem in a vertical format.

2. Multiply the number in the ones column by 3. $3 \times 4 = 12$
   Write the ones digit under the ones column as part of the product.
   Write the tens digit above the tens column.

$$\begin{array}{r} {}^{1} \\ 24 \\ \times\ 3 \\ \hline 2 \end{array}$$

3. Multiply 3 times the digit in the tens column. $3 \times 2 = 6$
   Add the 1 regrouped ten for a total of 7 tens.

$$\begin{array}{r} {}^{1} \\ 24 \\ \times\ 3 \\ \hline 72 \end{array}$$

4. Write the product. $3 \times 24 = 72$

YOUR TURN!

Find the product of 4 and 47.

1. Rewrite the problem in a vertical format.

2. Multiply the number in the ones column by 4.
   Write the ones digit under the ones column as part of the product.
   Write the tens digit above the tens column.

$$\begin{array}{r} 47 \\ \times\ 4 \\ \hline \end{array}$$

3. Multiply 4 times the digit in the tens column.

   $4 \times 4 =$ _____
   Add the 2 regrouped tens for a total of _____ tens.

$$\begin{array}{r} {}^{2} \\ 47 \\ \times\ 4 \\ \hline 8 \end{array}$$

4. Write the product. $4 \times 47 =$ _____

Who is Correct?

Find the product of 15 and 6.

Steve

15
×6
80

Allie

15
×6
60

Kyle

15
×6
90

Circle correct answer(s). Cross out incorrect answer(s).

 Guided Practice

Draw an array to model each expression. Find each product.

1 $3 \times 6 =$ _____

2 $2 \times 4 =$ _____

Step by Step Practice

3 Find the product of 57 and 6.

Step 1 Rewrite the problem in a vertical format.

Step 2 Multiply the number in the ones column by 6. $6 \times 7 =$ _____
Write the tens digit above the tens column.
Write the ones digit under the ones column as part of the product.

57
× 6

Step 3 Multiply 6 times the digit in the tens column.

$6 \times 5 =$ _____

Add the 4 regrouped tens for a total of _____ tens.

57
× 6

Step 4 Write the product.

$57 \times 6 =$ _____

GO ON

Find each product.

4   12
     × 2

5   9
     × 3

6   7
     × 4

7   51
     × 4

8   33
     × 6

9   21
     × 3

10   6
      × 6

11 25
     × 2

12 16
     × 3

Step by Step Problem-Solving Practice

Solve.

13 **TRAINS**   A cargo train has 19 connecting cars. With 4 wheels needed per car, how many wheels are there altogether?

Understand   Read the problem. Write what you know.

There are _____ wheels for each car.

There are _____ connecting cars.

Plan   Pick a strategy. One strategy is to use logical reasoning.

Solve   Think: 19 is close to 20.

4 × 20 = _____, so the product

should be close to _____.

_____ tires are needed to assemble the entire shipment.

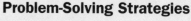

      19
     × 4

Check   Compare the answer to the estimate.

14 **FISH** Sea divers saw 6 schools of fish. Each school had 69 fish in each school. What is the total number of fish? Check off each step.

_____ Understand: I underlined key words.

_____ Plan: To solve the problem, I will _____.

_____ Solve: The answer is _____.

_____ Check: I checked my answer by _____.

FISH Sea divers saw 6 schools of fish.

15 **TEST SCORES** Mrs. Turner gives 4 points for each correct answer on the test. There are 65 questions. How many total possible points can each student earn?

_____

16 **Reflect** How is multiplication like repeated addition? Give an example and explain.

_____

_____

_____

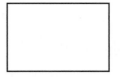

Skills, Concepts, and Problem Solving

Draw an array to model each expression. Find each product.

17 $6 \times 3 =$ _____

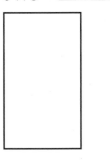

18 $3 \times 4 =$ _____

19 $4 \times 5 =$ _____

20 $2 \times 5 =$ _____

GO ON

Draw an array to model each expression. Find each product.

21 4 × 4 = _____

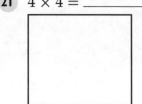

22 2 × 4 = _____

23 2 × 6 = _____

24 3 × 3 = _____

Find each product. Show your work.

25
  23
× 4

26
  63
× 6

27
  46
× 3

28
  98
× 2

29
  19
× 4

30
  73
× 3

31
  200
× 6

32
  55
× 2

33
  40
× 3

34
  493
× 4

35
  99
× 2

36
  222
× 6

37
  54
× 2

38
  107
× 4

39
  13
× 3

40
  75
× 3

41
  210
× 4

42
  316
× 2

43 BICYCLES   A bicycle shop has 234 bicycles in stock. Each bicycle has 2 wheels. How many wheels are in the bicycle shop?

_____

44 TREES   Each day, 315 trees are planted in Westin Woods. How many trees will be planted in 6 days?

_____

TREES 315 trees are planted in Westin Woods each day.

Vocabulary Check   **Write the vocabulary word that completes each sentence.**

45 _____ is an operation on two numbers to find their product.

46 A(n) _____ is objects or symbols displayed in rows of the same length and columns of the same length.

47 Writing in Math   Explain how to use a table of multiples to find 4×9.

_____

_____

Spiral Review (Lesson 5-1, p. 188)

Find each product.

48 $5 \times 1 =$ _____

49 $0 \times 26 =$ _____

50 $57 \times 10 =$ _____

51 $1 \times 85 =$ _____

52 $16 \times 0 =$ _____

53 $10 \times 91 =$ _____

54 $18 \times 5 =$ _____

55 $0 \times 108 =$ _____

Progress Check 1 (Lessons 5-1 and 5-2)

Find each product.

1. $3 \times 4 =$ _____

2. $4 \times 6 =$ _____

3. $6 \times 8 =$ _____

4. $5 \times 7 =$ _____

5. $10 \times 4 =$ _____

6. $2 \times 9 =$ _____

7.
$$\begin{array}{r} 98 \\ \times\ 1 \\ \hline \end{array}$$

8.
$$\begin{array}{r} 29 \\ \times\ 0 \\ \hline \end{array}$$

9.
$$\begin{array}{r} 41 \\ \times\ 5 \\ \hline \end{array}$$

10.
$$\begin{array}{r} 19 \\ \times\ 2 \\ \hline \end{array}$$

11.
$$\begin{array}{r} 10 \\ \times\ 17 \\ \hline \end{array}$$

12.
$$\begin{array}{r} 27 \\ \times\ 3 \\ \hline \end{array}$$

13.
$$\begin{array}{r} 71 \\ \times\ 6 \\ \hline \end{array}$$

14.
$$\begin{array}{r} 36 \\ \times\ 4 \\ \hline \end{array}$$

15.
$$\begin{array}{r} 92 \\ \times\ 10 \\ \hline \end{array}$$

Use the Identity Property of Multiplication to solve.

16. $32 \times$ _____ $=$ _____

17. _____ $\times 192 =$ _____

18. $15 \times$ _____ $= 150$

19. $10 \times 27 =$ _____

Use the Zero Property of Multiplication to solve.

20. $418 \times$ _____ $=$ _____

21. _____ $\times 22 =$ _____

Solve.

22. **LUNCH** Elliott eats 5 grapes every day for lunch. How many grapes will Elliott eat in 97 days?

_____

23. **SCHOOL** There are 23 students in class. How many fingers do these students have altogether?

_____

Multiply by 7, 8, and 9

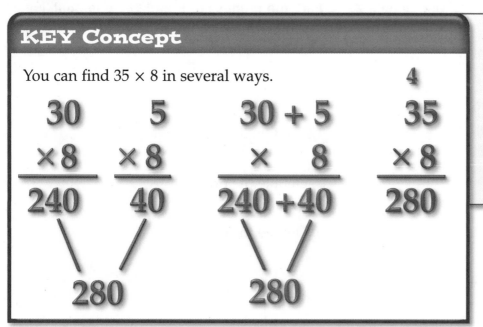

KEY Concept

You can find 35×8 in several ways.

$$\begin{array}{r} 30 \\ \times 8 \\ \hline 240 \end{array} \quad \begin{array}{r} 5 \\ \times 8 \\ \hline 40 \end{array} \quad \begin{array}{r} 30 + 5 \\ \times \quad 8 \\ \hline 240 + 40 \end{array} \quad \begin{array}{r} 4 \\ 35 \\ \times 8 \\ \hline 280 \end{array}$$

280   280

VOCABULARY

multiple
the product of the number and any whole number

product
the answer to a multiplication problem

Practice skip counting and memorizing the multiplication facts of 7, 8, and 9.

Example 1

Use a pattern to find the product of 9 × 5.

1. Skip count by 9s to write the multiples of 9. Facts with 9 make a pattern.

2. List the multiples of 9 up to ten:

   9, 18, 27, 36, 45, 54, 63, 72, 81, 90

3. Write the multiples as multiplication facts.

$1 \times 9 = 9$	$6 \times 9 = 54$
$2 \times 9 = 18$	$7 \times 9 = 63$
$3 \times 9 = 27$	$8 \times 9 = 72$
$4 \times 9 = 36$	$9 \times 9 = 81$
$5 \times 9 = 45$	$10 \times 9 = 90$

4. Write the product. $9 \times 5 = 45$

YOUR TURN!

Use a pattern to find the product of 6 × 7.

1. Skip count by 7s to write the multiples of 7.

2. List the multiples of 7 up to ten:

   _____, _____, _____, _____, _____,

   _____, _____, _____, _____, _____

3. Write the multiples as multiplication facts.

$1 \times 7 =$ _____	$6 \times 7 =$ _____
$2 \times 7 =$ _____	$7 \times 7 =$ _____
$3 \times 7 =$ _____	$8 \times 7 =$ _____
$4 \times 7 =$ _____	$9 \times 7 =$ _____
$5 \times 7 =$ _____	$10 \times 7 =$ _____

4. Write the product. $6 \times 7 =$ _____

GO ON

Example 2

Find the product of 8 and 21. Use doubling.

Double 1 and you have 2.
Double 2 and you have 4.
Double 4 and you have 8.
You can find the product of any number and 8 by doubling the number three times.

1. Double 21.

$$\begin{array}{r} \overset{1}{21} \\ \times\ 2 \\ \hline 42 \end{array}$$

2. Double the product.

$$\begin{array}{r} 42 \\ \times\ 2 \\ \hline 84 \end{array}$$

3. Double the product again.

$$\begin{array}{r} 84 \\ \times\ 2 \\ \hline 168 \end{array}$$

$8 \times 21 = 168$

YOUR TURN!

Find the product of 8 and 17. Use doubling.

Double 1 and you have 2.

Double 2 and you have _____.

Double 4 and you have _____.

1. Double _____.

$$\begin{array}{r} 17 \\ \times\ 2 \end{array}$$

2. Double the product.

$$\begin{array}{r} 34 \\ \times\ 2 \end{array}$$

3. Double the product again.

$$\begin{array}{r} 68 \\ \times\ 2 \end{array}$$

$8 \times 17 = $ _____

Who is Correct?

Find the product of 24 and 9.

Lurdes
$$\begin{array}{r} 34 \\ \times\ 9 \\ \hline 36 \\ +\ 27 \\ \hline 63 \end{array}$$

Nawat
$$\begin{array}{r} 34 \\ \times\ 9 \\ \hline 2,736 \end{array}$$

Kendra
$$\begin{array}{r} \overset{3}{34} \\ \times\ 9 \\ \hline 306 \end{array}$$

Circle correct answer(s). Cross out incorrect answer(s).

▶ Guided Practice

Use a pattern to find each product.

1 $7 \times 7 = $ _____

Multiples of 7: _____, _____, _____, _____, _____, _____, _____, _____, _____, _____

2 $6 \times 8 = $ _____

Multiples of 8: _____, _____, _____, _____, _____, _____, _____, _____, _____, _____

3 Find the product of 36 × 8. Use doubling.

   Step 1 Double 36.

   Step 2 Double the product.

   Step 3 Double the product again.

   Step 4 36 × 8 = _____

Find each product.

4 7 × 5 = _____

5 8 × 7 = _____

6 9 × 6 = _____

7 8 × 12 = _____

8 3 × 9 = _____

9 7 × 7 = _____

10
$$\begin{array}{r} 14 \\ \times\ 7 \\ \hline \end{array}$$

11
$$\begin{array}{r} 22 \\ \times\ 8 \\ \hline \end{array}$$

12
$$\begin{array}{r} 30 \\ \times\ 9 \\ \hline \end{array}$$

13
$$\begin{array}{r} 501 \\ \times\ \ \ 7 \\ \hline \end{array}$$

14
$$\begin{array}{r} 234 \\ \times\ \ \ 8 \\ \hline \end{array}$$

15
$$\begin{array}{r} 111 \\ \times\ \ \ 9 \\ \hline \end{array}$$

GO ON

Step by Step Problem-Solving Practice

Solve.

16 **TENNIS** Diana and her teammates are getting ready for the tennis tournament. There are 14 teams in the league, each with 7 players. How many players are in the league?

Understand Read the problem. Write what you know.

There are _____ tennis teams.

There are _____ players on each team.

Plan Pick a strategy. One strategy is to make a table. Make a table with two rows. Title one row "teams" and the other row "players."

Solve Write 1 through 14 for teams, because there are 14 teams in the league. There are 7 players on each team, so add 7 each time.

Teams	1	2	3	4	5	6								
Players	7	14	21											

There are _____ players in the league.

Check Multiply 14 × 7 to check your answer.

17 **SCHOOL** The lunch room at North School has 9 tables. Eight students can sit at each table. If all the seats are filled, how many students will be sitting in the lunch room? Check off each step.

_____ Understand: I underlined key words.

_____ Plan: To solve this problem, I will _____.

_____ Solve: The answer is _____.

_____ Check: I checked my answer by _____.

18 **WALKING** Andrew walks 4 miles per hour. If Andrew walks for 7 hours, how far will he have walked? Show your work.

_____

19 **Reflect** Explain how when multiplying by eight you can use doubles.

_____

_____

▶ Skills, Concepts, and Problem Solving

Use patterns to find each product.

20 $7 \times 10 =$ _____

Multiples of 7: _____, _____, _____, _____, _____, _____, _____, _____, _____, _____

21 $8 \times 8 =$ _____

Multiples of 8: _____, _____, _____, _____, _____, _____, _____, _____, _____, _____

22 $9 \times 9 =$ _____

Multiples of 9: _____, _____, _____, _____, _____, _____, _____, _____, _____, _____

Find each product.

23 $2 \times 7 =$ _____

24 $6 \times 8 =$ _____

25 $3 \times 9 =$ _____

26 $7 \times 3 =$ _____

27 $8 \times 5 =$ _____

28 $9 \times 4 =$ _____

29 $\begin{array}{r} 11 \\ \times\ 7 \\ \hline \end{array}$

30 $\begin{array}{r} 20 \\ \times\ 9 \\ \hline \end{array}$

31 $\begin{array}{r} 26 \\ \times\ 8 \\ \hline \end{array}$

32 $\begin{array}{r} 29 \\ \times\ 7 \\ \hline \end{array}$

33 $\begin{array}{r} 49 \\ \times\ 8 \\ \hline \end{array}$

34 $\begin{array}{r} 99 \\ \times\ 9 \\ \hline \end{array}$

35 $\begin{array}{r} 211 \\ \times\ 7 \\ \hline \end{array}$

36 $\begin{array}{r} 880 \\ \times\ 8 \\ \hline \end{array}$

37 $\begin{array}{r} 109 \\ \times\ 6 \\ \hline \end{array}$

GO ON

38 **EARNINGS** Roberto earns $8 an hour delivering newspapers. If Roberto works for 12 hours a week, how much will he have earned?

_____

39 **BAKING** Al's bakery has 35 baskets full of rolls. Each basket has 9 rolls inside. How many rolls are there altogether?

_____

EARNINGS Roberto earns $8 an hour delivering newspapers.

Vocabulary Check **Write the vocabulary word or words that complete each sentence.**

40 A(n) _____ of a number is the product of that number and any whole number.

41 The answer or result of a multiplication problem is called the

_____.

42 **Writing in Math** Write the multiples of 9 from 1 to 10. Explain any pattern in the products.

_____

_____

_____

 Spiral Review (Lessons 5-1, p. 188 and 5-2, p. 195)

Find each product.

43 5 × 44 = _____

44 0 × 999 = _____

45 92 × 3 = _____

Solve. (Lesson 5-2, p. 195)

46 **SUPPLIES** Alexis bought 6 packs of pens. There are 15 pens in each pack. How many pens did Alexis buy?

_____

 STOP

Multiply by 11 and 12

KEY Concept

When multiplying by 11 and 12, you can use **patterns** and **array** models.

Multiply by 11		
1	× **11** =	11
2	× **11** =	22
3	× **11** =	33
4	× **11** =	44
5	× **11** =	55
6	× **11** =	66
7	× **11** =	77
8	× **11** =	88
9	× **11** =	99

The table shows that when a single-digit number is multiplied by 11, the **product** is the digit repeated.

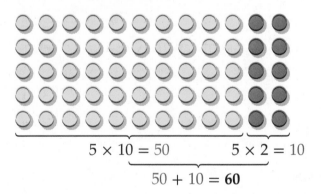

$5 \times 10 = 50 \qquad 5 \times 2 = 10$

$50 + 10 = \mathbf{60}$

The array model shows that you can think of 12×5 as $(10 \times 5) + (2 \times 5)$. This is the **Distributive Property of Multiplication**.

VOCABULARY

array
    objects or symbols in rows of the same length and columns of the same length; the length of a row might be different from the length of a column

Distributive Property of Multiplication
    to multiply a sum by a number, multiply each addend by the number outside the parenthesis

pattern(s)
    a sequence of numbers, figures, or symbols that follows a rule or design

product
    the answer to a multiplication problem

Use array models, patterns, or the Distributive Property of Multiplication to multiply by 11 and 12.

GO ON

Example 1

Use a pattern to find 7 × 11.

1. Use a table to complete the pattern.

1	× 11 =	11
2	× 11 =	22
3	× 11 =	33
4	× 11 =	44
5	× 11 =	55
6	× 11 =	66
7	× 11 =	77

2. Write the product.   7 × 11 = 77

YOUR TURN!

Use a pattern to find 7 × 12.

1. Use a table to complete the pattern.

1	× 12 =	12
2	× 12 =	24
3	× 12 =	36
4	× 12 =	48
5		
6		
7		

2. Write the product.   7 × 12 = _____

Example 2

Use an array model to find 12 × 7.

1. Rewrite 12 × 7 as (**10** × 7) + (**2** × 7).
   Then use the Distributive Property.

2. Draw an array using two sets of counters.

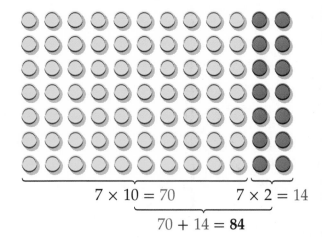

   7 × 10 = 70   7 × 2 = 14
              70 + 14 = **84**

3. The array shows 10 × 7 = 70
   and 2 × 7 = **14**.
   So, 70 + **14** = **84**.

4. Write the product.   12 × 7 = **84**

YOUR TURN!

Use an array model to find 12 × 4.

1. Rewrite 12 × 4.

   (_____ × _____) + (_____ × _____)

2. Draw an array using two sets of counters.

3. The array shows 10 × _____ = _____

   and 2 × _____ = _____.

   So, _____ + _____ = _____.

4. Write the product.   12 × 4 = _____

Who is Correct?

Find the product of 11 and 12.

Nicole
(11 × 12) =
(10 × 11) + (2 × 11)
= 110 + 22
= 132

Pablo
10 + 2
×10 + 1
100 + 2
= 102

Tania
11 × 12 =
(10 × 10) + (1 × 12)
= 100 + 12
= 112

Circle correct answer(s). Cross out incorrect answer(s).

 Guided Practice

Rewrite the factors using the Distributive Property.

1   12 × 6

_____ = (___ × ___) + (___ × ___)

2   12 × 3

_____ = (___ × ___) + (___ × ___)

Step by Step Practice

3   Use an array model to find 11 × 4.

Step 1   Rewrite 11 × 5.

(_____ × _____) + (1 × _____)

Step 2   Draw an array using two sets of counters.

Step 3   11 × 4 = (☐ × ☐) + (☐ × ☐)

= ☐ + ☐

= ☐

GO ON

Use a pattern or draw an array model to find each product.

4 $11 \times 12 =$ _____

5 $12 \times 9 =$ _____

Find each product.

6 $12 \times 10 =$ _____

7 $11 \times 7 =$ _____

8 $8 \times 11 =$ _____

9 $6 \times 12 =$ _____

Step by Step *Problem-Solving Practice*

Solve.

10 REMODELING   Enrique is helping to place new tile on his kitchen floor. He is responsible for 12 feet by 10 feet of the floor. If each tile is 1 foot, how many tiles does Enrique need for the kitchen floor?

Problem-Solving Strategies
☐ Draw a picture.
☐ Use logical reasoning.
☑ Solve a simpler problem.
☐ Work backward.
☐ Make a table.

Understand   Read the problem. Write what you know.

Enrique is responsible for _____ by _____ feet of the kitchen floor.

Each tile is _____ foot.

Plan   Pick a strategy. One strategy is solve a simpler problem.

Solve   Write 12×10 using the Distributive Property.

(_____ × _____) + (_____ × _____)

Add the products. _____

Enrique needs _____ tiles.

Check   You can skip count by 10s to the 12th number.

11 **AUDITORIUM**   There are 11 seats in each row in the school auditorium. If the sixth grade class takes up 7 rows, how many students are there in all?
Check off each step.

_____ **Understand: I underlined key words.**

_____ **Plan: To solve the problem, I will** _____.

_____ **Solve: The answer is** _____.

_____ **Check: I checked my answer by** _____.

12 **Reflect**   Why does the pattern for multiplying by 11 only work for single digits?

_____

_____

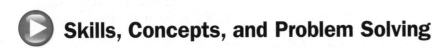 ## Skills, Concepts, and Problem Solving

Use patterns or array models to find each product.

13 $11 \times 6 =$ _____

14 $12 \times 5 =$ _____

15 $12 \times 7 =$ _____

16 $11 \times 7 =$ _____

GO ON

Find each product.

17 $11 \times 6 =$ _____

18 $12 \times 4 =$ _____

19 $11 \times 11 =$ _____

20 $12 \times 3 =$ _____

21 $12 \times 11 =$ _____

22 $1 \times 11 =$ _____

23 $12 \times 2 =$ _____

24 $11 \times 10 =$ _____

26 $6 \times 12 =$ _____

Solve.

26 HIKING   The hikers plan to travel 12 miles each day for 7 days. How many miles do they plan on traveling in all?

_____

27 PHOTOS   Tomas placed 9 photos on 12 photo album pages. How many photos did he place altogether?

_____

28 EARNINGS   Melissa earns $11 an hour. How much will she earn if she works 9 hours in one week?

_____

HIKING   The hikers will travel 12 miles each day for 7 days.

Vocabulary Check   **Write the vocabulary word that completes each sentence.**

29 A(n) _____ follows a rule or design.

30 Writing in Math   Explain how the word *dozen* can help you multiply by 12.

_____

_____

 Spiral Review

Find each product. (Lesson 5-3, p. 203)

31 $7 \times 4 =$ _____

32 $8 \times 9 =$ _____

33 $9 \times 5 =$ _____

Solve. (Lesson 5-1, p. 188)

34 FLOWERS   Belinda is making bunches of flowers for her mother, her three aunts, and her grandfather. Each bunch will have 10 flowers. How many flowers will Belinda need to make 5 bunches?

_____

STOP

Multiply Greater Numbers

KEY Concept

Traditional Multiplication Method

$$
\begin{array}{r}
{\scriptstyle 1} \\
43 \\
\times\ 35 \\
\hline
215 \\
+\ 1\,290 \\
\hline
1{,}505
\end{array}
$$

Partial Products Method

$$
\begin{array}{r}
43 \\
\times\ 35 \\
\hline
15 \\
200 \\
90 \\
+\ 1{,}200 \\
\hline
1{,}505
\end{array}
$$

$5 \times 3 = 15$
$5 \times 40 = 200$
$30 \times 3 = 90$
$30 \times 40 = 1{,}200$

VOCABULARY

estimate
a number close to an exact value; an estimate indicates *about* how much

partial products method
a way to multiply; the value of each digit in one factor is multiplied by the value of each digit in the other factor; the product is the sum of its partial products

Before you multiply, you should **estimate** your answer.
Then check your actual answer for reasonableness.

Example 1

Find the product of 37 and 62. Use the partial products method.

1. Estimate.   **40 × 60 = 2,400**

2. Rewrite the problem in a vertical format.

3. Multiply 2 times the ones column.
   2 × 7 = 14

4. Multiply 2 times the tens column.
   2 × 30 = 60

5. Multiply 60 times the ones column.
   60 × 7 = 420

6. Multiply 60 times the tens column.
   60 × 30 = 1,800

$$
\begin{array}{r}
37 \\
\times\ 62 \\
\hline
14 \\
60 \\
420 \\
+1{,}800 \\
\hline
2{,}294
\end{array}
$$

7. Add the partial products.

   14 + 60 + 420 + 1,800 = 2,294

8. 37 × 62 = 2,294

YOUR TURN!

Find the product of 38 and 47. Use the partial products method.

1. Estimate. 40 × 50 = _____

2. Rewrite the problem in a vertical format.

3. Multiply _____ times the ones column.
   7 × 8 = _____

4. Multiply _____ times the tens column.
   7 × 30 = _____

5. Multiply 40 times the ones column.
   40 × 8 = _____

6. Multiply 40 times the tens column.
   40 × 30 = _____

$$
\begin{array}{r}
38 \\
\times\ 47 \\
\end{array}
$$

7. Add the partial products.

   _____ + _____ + _____ + _____

   = _____

8. 38 × 47 = _____

GO ON

Example 2

Find the product of 56 and 14. Use the traditional multiplication method.

1. Estimate.
   60 × 10 = 600

2. Rewrite the problem in a vertical format.

3. Multiply 4 times the digit in the ones column.
   4 × 6 = 24
   Write the tens digit above the tens column. Write the ones digit in the product under the ones column.

$$\begin{array}{r} \overset{2}{56} \\ \times\ 14 \\ \hline 4 \end{array}$$

4. Multiply 4 times the tens column.
   4 × 5 = 20
   Add the two tens to get 22.

$$\begin{array}{r} \overset{2}{56} \\ \times\ 14 \\ \hline 224 \end{array}$$

5. Multiply the value of the digit in the tens place times 6.
   10 × 6 = 60

$$\begin{array}{r} 56 \\ \times\ 14 \\ \hline 224 \\ 60 \end{array}$$

6. Multiply the value in the tens column by 10.
   10 × 50 = 500
   Write the 5 in the hundreds place.

$$\begin{array}{r} 56 \\ \times\ 14 \\ \hline 224 \\ +\ 560 \\ \hline 784 \end{array}$$

7. Find the sum of the two products.

8. 56 × 14 = 784

   Compare to your estimate for reasonableness.

YOUR TURN!

Find the product of 78 and 32. Use the traditional multiplication method.

1. Estimate.

   _____ × _____ = _____

2. Rewrite the problem in a vertical format.

3. Multiply _____ times the digit in the ones column.
   2 × 8 = _____

$$\begin{array}{r} \overset{1}{78} \\ \times\ 32 \\ \hline \end{array}$$

4. Multiply _____ times the tens column.

   2 × 7 = _____

   Add the _____ ten(s)

   to get _____.

5. Multiply each place value by the tens digit.

   30 × 8 = _____

6. Multiply the value in the tens column by 30.

   30 × 70 = _____

   Add _____ hundred(s) for

   _____.

7. Find the sum of the two products.

8. 78 × 32 = _____

   Compare to your estimate for reasonableness.

Who is Correct?

Find the product of 18 and 53.

Vivian

18
× 53
24
3
40
+ 5
72

Alonso

18
× 53
24
30
400
+ 500
954

Larry

18(50 + 3)
= (18 × 50) + (18 × 3)
= 900 + 54
= 954

Circle correct answer(s). Cross out incorrect answer(s).

▶ Guided Practice

Estimate each product.

1 70 × 32

Round each factor to the greatest place value.
Find the estimated product.

_____ × _____ = _____

2 95 × 47

Round each factor to the greatest place value.
Find the estimated product.

_____ × _____ = _____

Step (by) Step **Practice**

3 Find the product of 13 and 29. Use the traditional multiplication method.

Step 1 Rewrite the problem in a vertical format.

Step 2 Multiply each place value by the ones digit.

13
× 29

Step 3 Multiply each place value by the tens digit.

Step 4 Find the sum of the two products.

Step 5 13 × 29 = _____

GO ON

Find each product. Use the traditional multiplication method.

4
$$\begin{array}{r} 11 \\ \times\,71 \\ \hline \end{array}$$

5
$$\begin{array}{r} 12 \\ \times\,44 \\ \hline \end{array}$$

Find each product. Use the partial products method.

6
$$\begin{array}{r} 39 \\ \times\,83 \\ \hline \end{array}$$

7
$$\begin{array}{r} 28 \\ \times\,94 \\ \hline \end{array}$$

Step by Step *Problem-Solving Practice*

Solve.

8 **TRAVEL** Mr. Simmons drives from Tampa to Orlando 25 times a year. The distance between the two cities is 85 miles. Each drive is a round trip. How many total miles does Mr. Simmons travel in a year?

Problem-Solving Strategies
☐ Draw a diagram.
☑ Use logical reasoning.
☐ Solve a simpler problem.
☐ Work backward.
☐ Make a table.

Understand Read the problem. Write what you know.
Mr. Simmons drives from Tampa to Orlando

_____ times a year.

The distance between the cities is _____ miles.

Each drive is a _____ trip.

Plan Pick a strategy. One strategy is to use logical reasoning.

Solve distance between cities × number of times traveled = one way total miles traveled

one way total miles traveled × round trip = total miles traveled

Use the values from the problem to solve.

_____ × _____ = _____

_____ × _____ = _____

Mr. Simmons travels _____ miles in a year.

Check Use estimation to check.

9 BUSINESS Greg is digging for clams to sell to a local seafood store. For every shovel of sand he digs up, he finds 16 small clams. Greg did this for 3 days. If he digs 27 shovels of sand, how many clams could he expect to find?
Check off each step.

_____ Understand: I underlined key words.

_____ Plan: To solve the problem, I will _____.

_____ Solve: The answer is _____.

_____ Check: I checked my answer by _____.

10 COOKING When making meat loaf, Ryan uses 12 ounces of breadcrumbs. His mother uses the same recipe at her restaurant, but she multiplies the recipe to make 17 meat loaves. How many ounces of breadcrumbs will his mother use?

_____

11 **Reflect** Compare two different ways of multiplying 13 × 17.

_____

_____

 # Skills, Concepts, and Problem Solving

Find each product. Use the traditional multiplication method.

12
$$\begin{array}{r} 56 \\ \times\, 91 \\ \hline \end{array}$$

13
$$\begin{array}{r} 45 \\ \times\, 87 \\ \hline \end{array}$$

14
$$\begin{array}{r} 13 \\ \times\, 14 \\ \hline \end{array}$$

15
$$\begin{array}{r} 11 \\ \times\, 15 \\ \hline \end{array}$$

GO ON

Find each product. Use the partial products method.

16 39
      × 63

17 18
      × 22

18 23
      × 28

19 32
      × 16

20 32
      × 31

21 67
      × 14

Find each product. Show your work.

22 26
      × 31

23 44
      × 62

24 15
      × 48

25 17
      × 45

26 51
      × 23

27 29
      × 33

Solve.

28 **COOKING**   Tierra and her sister are making sandwiches for their day camp. Each loaf of bread has 16 slices. They use 14 loaves. How many slices of bread did they use in all?

_____

29 **PHOTOGRAPHY**   Tanya's online photo album holds 50 pictures. She has 18 photo albums. How many pictures does she have in all?

_____

COOKING Tierra and her sister are making sandwiches.

Solve.

30 **TRIP** Paul and his family are taking a 14-day road trip. They plan to drive 130 miles each day. How many miles would this be?

_____

31 **MUSIC** Martin has 63 CDs. If each CD has 15 songs on it, how many total songs does Martin have?

_____

TRIP Paul and his family are taking a 14-day road trip.

Vocabulary Check **Write the vocabulary word that completes each sentence.**

32 A number close to an exact value is called a(n) _____.

33 **Writing in Math** Shaun multiplied 45 × 18. What mistake did he make?

_____

_____

_____

$$\begin{array}{r} 4 \\ 45 \\ \times\ 18 \\ \hline 360 \\ +345 \\ \hline 705 \end{array}$$

 Spiral Review

Solve. (Lesson 5-3, p. 203)

34 **GROCERY** There are 7 egg cartons on the shelf at the corner market. Each egg carton contains 1 dozen eggs. What is the total number of eggs on the shelf?

_____

Find each product. (Lesson 5-4, p. 209)

35 11 × 8 = _____

36 12 × 7 = _____

37 12 × 13 = _____

38 11 × 46 = _____

39 12 × 31 = _____

40 11 × 52 = _____

STOP

Progress Check 2 (Lessons 5-3, 5-4, and 5-5)

Draw an array to model each expression.

1 $7 \times 3 =$ _____

2 $8 \times 6 =$ _____

Find each product. Use the Distributive Property.

3 $5 \times 12 = ($ _____ $\times 10) + ($ _____ $\times 2)$

   $=$ _____ $+$ _____

   $=$ _____

4 $12 \times 7 = ($ _____ $\times 7) + ($ _____ $\times 7)$

   $=$ _____ $+$ _____

   $=$ _____

5 $11 \times 4 = ($ _____ $\times 4) + ($ _____ $\times 4)$

   $=$ _____ $+$ _____

   $=$ _____

6 $11 \times 8 = (10 \times$ _____ $) + (1 \times$ _____ $)$

   $=$ _____ $+$ _____

   $=$ _____

Find each product. Show your work.

7 $6 \times 11 =$ _____

8 $8 \times 9 =$ _____

9 $7 \times 4 =$ _____

10 $5 \times 9 =$ _____

11 $8 \times 12 =$ _____

12 $11 \times 2 =$ _____

13
$$\begin{array}{r} 24 \\ \times\, 68 \\ \hline \end{array}$$

14
$$\begin{array}{r} 87 \\ \times\, 19 \\ \hline \end{array}$$

15
$$\begin{array}{r} 46 \\ \times\, 72 \\ \hline \end{array}$$

Solve.

16 **MUSEUM**   On Saturday 12 school groups visited the museum. Each group had 20 students. How many students visited the museum on Saturday?

_____

17 **SUPPLIES**   Olivia bought 25 packs of pencils. There are 12 pencils in each pack. How many pencils did Olivia buy?

_____

Vocabulary and Concept Check

array, *p. 195*

Distributive Property
of Multiplication, *p. 209*

estimate, *p. 215*

factor, *p. 188*

Identity Property of
Multiplication, *p. 188*

multiple, *p. 195*

multiplication, *p. 188*

partial products
method, *p. 215*

pattern(s), *p. 209*

product, *p. 188*

Zero Property of
Multiplication, *p. 188*

Write the vocabulary word that completes each sentence.

1 A number that is multiplied by another number to find a

product is called a(n) _____.

2 The _____ states that any
number multiplied by zero is zero.

3 _____ is an operation on two
numbers to find their product.

4 A number close to an exact value is a(n)

_____.

5 A sequence of numbers, figures, or symbols that follow a rule

or design is called a(n) _____.

Label each diagram below. Write the correct vocabulary term in each blank.

6 _____

$$4: 4, 8, 12, 16, 20, 24$$

7 _____

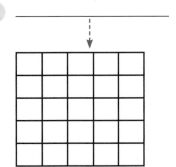

8 _____

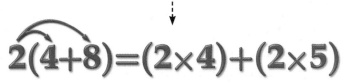

$$2(4+8)=(2\times4)+(2\times5)$$

Lesson Review

5-1 Multiply by 0, 1, 5, and 10 (pp. 188–194)

Find each product.

9 $16 \times 0 =$ _____

10 $92 \times 10 =$ _____

11 $5 \times 15 =$ _____

12 $1 \times 310 =$ _____

13 $0 \times 402 =$ _____

14 $71 \times 10 =$ _____

15 $48 \times 1 =$ _____

16 $20 \times 5 =$ _____

Example 1

Find the product of 5 and 13.

1. Rewrite the problem in a vertical format.

2. Multiply the number in the ones column by 5.
   $3 \times 5 = 15$
   Write the tens digit above the tens column. Write the ones digit under the ones column.

   $$\begin{array}{r} \overset{1}{13} \\ \times\ 5 \\ \hline 5 \end{array}$$

3. Multiply 5 times the digit in the tens column.
   $5 \times 1 = 5$
   Add one regrouped tens for a total of 6 tens.

   $$\begin{array}{r} \overset{1}{13} \\ \times\ 5 \\ \hline 65 \end{array}$$

4. $5 \times 13 = 65$

5-2 Multiply by 2, 3, 4, and 6 (pp. 195–201)

Find each product.

17 $17 \times 2 =$ _____

18 $12 \times 6 =$ _____

19 $4 \times 30 =$ _____

20 $21 \times 3 =$ _____

21 $2 \times 28 =$ _____

22 $3 \times 17 =$ _____

23 $51 \times 4 =$ _____

Example 2

Find the product of 4 and 32.

1. Rewrite the problem in a vertical format.

2. Multiply the number in the ones column by 4. $4 \times 2 = 8$
   Write the ones digit under the ones column.

   $$\begin{array}{r} 32 \\ \times\ 4 \\ \hline 8 \end{array}$$

3. Multiply 4 times the digit in the tens column.
   $4 \times 3 = 12$
   Write the tens digits under the tens column.

   $$\begin{array}{r} 32 \\ \times\ 4 \\ \hline 128 \end{array}$$

4. $4 \times 32 = 128$

5-3 Multiply by 7, 8, and 9 (pp. 203–208)

Find each product.

24. $7 \times 16 =$ _____

25. $8 \times 8 =$ _____

26. $9 \times 7 =$ _____

27. $1 \times 8 =$ _____

28. $6 \times 9 =$ _____

29. $9 \times 7 =$ _____

30. $8 \times 5 =$ _____

31. $9 \times 12 =$ _____

32.
$$\begin{array}{r} 31 \\ \times\ 8 \\ \hline \end{array}$$

33.
$$\begin{array}{r} 56 \\ \times\ 9 \\ \hline \end{array}$$

34.
$$\begin{array}{r} 19 \\ \times\ 7 \\ \hline \end{array}$$

35.
$$\begin{array}{r} 18 \\ \times\ 8 \\ \hline \end{array}$$

36.
$$\begin{array}{r} 12 \\ \times\ 9 \\ \hline \end{array}$$

37.
$$\begin{array}{r} 23 \\ \times\ 7 \\ \hline \end{array}$$

Example 3

Use a pattern to find the product of 9 × 6.

1. Skip count by 9s to write the multiples of 8.

2. List the multiples of 9 up to ten:
   9, 18, 27, 36, 45, 54, 63, 72, 81, 90

3. Write the multiples as multiplication facts.

$1 \times 9 = 9$	$6 \times 9 = 54$
$2 \times 9 = 18$	$7 \times 9 = 63$
$3 \times 9 = 27$	$8 \times 9 = 72$
$4 \times 9 = 36$	$9 \times 9 = 81$
$5 \times 9 = 45$	$10 \times 9 = 90$

Example 4

Find the product of 8 and 35. Use doubling.

Double 1 and you have 2.

Double 2 and you have 4.

Double 4 and you have 8.

$$\begin{array}{r} 35 \\ \times\ 2 \\ \hline 70 \end{array}$$

You can find the product of any number and 8 by doubling the number 3 times.

$$\begin{array}{r} 70 \\ \times\ 2 \\ \hline 140 \end{array}$$

1. Double 35.

2. Double the product.

$$\begin{array}{r} 140 \\ \times\ 2 \\ \hline 280 \end{array}$$

3. Double the product again.

$8 \times 35 = 280$

5-4 Multiply by 11 and 12 (pp. 209–214)

Find each product.

38 $4 \times 11 =$ _____

39 $12 \times 6 =$ _____

40 $11 \times 12 =$ _____

41 $3 \times 12 =$ _____

42 $11 \times 8 =$ _____

43 $12 \times 7 =$ _____

44 $3 \times 11 =$ _____

45 $9 \times 12 =$ _____

46 $11 \times 5 =$ _____

47 $11 \times 12 =$ _____

48 $12 \times 10 =$ _____

49 $4 \times 12 =$ _____

50 $9 \times 11 =$ _____

51 $12 \times 5 =$ _____

Example 5

Use an array model to find 12×8.

1. Rewrite 12×8 in distributive form.
   $(10 \times 8) + (2 \times 8)$

2. Draw an array using two sets of counters.

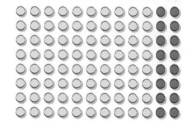

3. The array shows $10 \times 8 = 80$ and
   $2 \times 8 = 16$. So, $80 + 16 = 96$.

4. $12 \times 8 = 96$

Example 6

Use a pattern to find 8×11.

1. Use a table to complete the pattern.

1	$\times \mathbf{11} =$	11
2	$\times \mathbf{11} =$	22
3	$\times \mathbf{11} =$	33
4	$\times \mathbf{11} =$	44
5	$\times \mathbf{11} =$	55
6	$\times \mathbf{11} =$	66
7	$\times \mathbf{11} =$	77
8	$\times \mathbf{11} =$	88

2. $8 \times 11 = 88$

5-5 Multiply Large Numbers (pp. 215–221)

Find each product. Use the partial products method.

52 $\begin{array}{r} 13 \\ \times\ 18 \\ \hline \end{array}$

53 $\begin{array}{r} 47 \\ \times\ 62 \\ \hline \end{array}$

54 $\begin{array}{r} 26 \\ \times\ 53 \\ \hline \end{array}$

55 $\begin{array}{r} 71 \\ \times\ 29 \\ \hline \end{array}$

Example 7

Find the product of 31 and 42. Use the partial products method.

1. Rewrite the problem in a vertical format.

2. Multiply 2 times the ones column.
   $2 \times 1 = 2$

3. Multiply 2 times the tens column.
   $2 \times 30 = 60$

4. Multiply 40 times the ones column.
   $40 \times 1 = 40$

5. Multiply 40 times the tens column.
   $40 \times 30 = 1{,}200$

6. Add the partial products.
   $2 + 60 + 40 + 1{,}200 = 1{,}302$

7. $31 \times 42 = 1{,}302$

$\begin{array}{r} 31 \\ \times\ 42 \\ \hline 2 \\ 60 \\ 40 \\ 1{,}200 \\ \hline 1{,}302 \end{array}$

Find each product. Use the traditional multiplication method.

56 $\begin{array}{r} 80 \\ \times\ 76 \\ \hline \end{array}$

57 $\begin{array}{r} 43 \\ \times\ 12 \\ \hline \end{array}$

58 $\begin{array}{r} 28 \\ \times\ 14 \\ \hline \end{array}$

59 $\begin{array}{r} 39 \\ \times\ 34 \\ \hline \end{array}$

60 $\begin{array}{r} 51 \\ \times\ 82 \\ \hline \end{array}$

61 $\begin{array}{r} 65 \\ \times\ 43 \\ \hline \end{array}$

Chapter Test

Draw an array to model each expression.

1 $9 \times 3 =$ _____

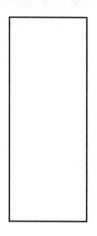

2 $8 \times 8 =$ _____

Find each product.

3 $0 \times 481 =$ _____

4 $97 \times 1 =$ _____

5 $7 \times 5 =$ _____

6 $18 \times 10 =$ _____

7 $13 \times 2 =$ _____

8 $3 \times 8 =$ _____

9
$$\begin{array}{r} 21 \\ \times\,14 \\ \hline \end{array}$$

10
$$\begin{array}{r} 14 \\ \times\,6 \\ \hline \end{array}$$

11
$$\begin{array}{r} 16 \\ \times\,47 \\ \hline \end{array}$$

12
$$\begin{array}{r} 28 \\ \times\,28 \\ \hline \end{array}$$

13
$$\begin{array}{r} 20 \\ \times\,9 \\ \hline \end{array}$$

14
$$\begin{array}{r} 11 \\ \times\,13 \\ \hline \end{array}$$

15
$$\begin{array}{r} 12 \\ \times\,84 \\ \hline \end{array}$$

16
$$\begin{array}{r} 17 \\ \times\,25 \\ \hline \end{array}$$

17
$$\begin{array}{r} 12 \\ \times\,8 \\ \hline \end{array}$$

Use partial products to find each product.

18
```
    29
  × 38
```

19
```
    44
  × 57
```

Solve.

20 **FRUIT** Brandy is picking strawberries from his garden to sell at the market. He has 35 strawberries in 16 baskets. How many strawberries does Brandy have in all?

_____

21 **FITNESS** Ron is lifting weights at the gym. Ron has 4 weights on each side of the barbell. If each weight weighs 10 pounds, how much is he lifting?

_____

Correct the mistakes.

22 **SNACKS** Courtney wanted to share raisins with Norman and Mark. She counted 36 raisins in his box. She said, "If we all have the same number of raisins, there are 98 raisins in all." What is wrong with Courtney's statement?

_____

_____

STOP

Choose the best answer and fill in the corresponding circle on the sheet at right.

1 $36 \times 1 = 36$ is an example of which property?

   A Distributive Property

   B Zero Property of Multiplication

   C Identity Property of Multiplication

   D Commutative Property of Multiplication

2 $68 \times 2 =$

   A 34   **C** 136

   B 70   **D** 163

3 Find the product of 12 and 52.

   A 624   **C** 246

   B 64   **D** 46

4 A set of bleachers has 14 rows of seats. Each row can seat 35 people. If the bleachers are full, how many people are seated on the bleachers?

   A 21

   B 49

   C 390

   D 490

5 The first four multiples of 9 are…

   A 1, 3, 3, 9   **C** 36

   B 9, 18, 27, 36   **D** 9, 10, 11, 12

6 Gloria cut a rope into 7 pieces. Each piece is 13 ft long. How long was the original rope?

   A 21 feet   **C** 98 feet

   B 91 feet   **D** 104 feet

7 Which expression does this array model?

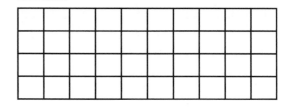

   A 10×1 **C** 4×10

   B 40   **D** 40×4

8 Jaden is walking dogs for his dog walking business. If he counts 9 dogs legs, how many paws are there in all?

   A 36   **C** 42

   B 32   **D** 44

9 $86 \times 7 =$

   A 93   **C** 522

   B 206   **D** 602

10 Each day there are 7 tours at the aquarium. Twenty-eight people can go on a tour. How many people can tour the aquarium each day?

    A 156

    B 196

    C 200

    D 216

11 The cafeteria has 5 long tables. Each table can seat 10 students. How many students can sit at all of the tables?

    A 2

    B 20

    C 50

    D 55

12 The product of 77 and 7 is _____.

    A 11

    B 70

    C 84

    D 539

ANSWER SHEET

Directions: Fill in the circle of each correct answer.

1. (A) (B) (C) (D)
2. (A) (B) (C) (D)
3. (A) (B) (C) (D)
4. (A) (B) (C) (D)
5. (A) (B) (C) (D)
6. (A) (B) (C) (D)
7. (A) (B) (C) (D)
8. (A) (B) (C) (D)
9. (A) (B) (C) (D)
10. (A) (B) (C) (D)
11. (A) (B) (C) (D)
12. (A) (B) (C) (D)

Success Strategy

Double check your answers after you finish. Read each problem and all of the answer choices. Put your finger on each bubble you filled in to make sure it matches the answer for each problem.

STOP

Chapter 6

Division

Soccer players are divided into teams.

Division helps you group things together, or find out how many groups there can be. For example, division can help you determine teams. If there are 40 people that want to play soccer, how many people will be on each of the 4 teams?

STEP **1** Quiz

Math Online ⟩ Are you ready for Chapter 6? Take the Online Readiness Quiz at *glencoe.com* to find out.

STEP **2** Preview

Get ready for Chapter 6. Review these skills and compare them with what you will learn in this chapter.

What You Know	What You Will Learn

What You Know

You know that there are properties and rules when you multiply.

Identity Property of Multiplication

$14 \times 1 = 14$ $22 \times 1 = 22$

Zero Property of Multiplication

$30 \times 0 = 0$ $210 \times 0 = 0$

What You Will Learn

Lesson 6-1

There are special division rules when you divide.

Any number divided by itself is equal to 1. $3 \div 3 = 1$

You cannot divide by 0. It is not possible.

You know that addition and subtraction are inverse operations.

Example: $5 + 7 = 12$
            $12 - 7 = 5$

TRY IT!

Rewrite each equation using an inverse operation.

1 $9 - 5 = 4$

   _____

2 $15 + 10 = 25$

   _____

3 $64 - 31 = 33$

   _____

4 $22 + 76 = 98$

   _____

Lessons 6-2 through 6-6

Multiplication and division are **inverse operations**.

$$\boxed{5} \times \boxed{7} = \boxed{35}$$
$$\boxed{35} \div \boxed{7} = \boxed{5}$$

$$\boxed{9} \times \boxed{3} = \boxed{27}$$
$$\boxed{27} \div \boxed{3} = \boxed{9}$$

Division with 0, 1, and 10

KEY Concept

There are special division rules to use when you divide.

Any number divided by itself is equal to 1.

$$4 \div 4 = 1$$

Any number divided by 1 is the same number.

$$4 \div 1 = 4$$

Zero divided by any number (except 0) equals 0.

You cannot divide by 0. It is not possible.

You can use models to divide by ten.

$$40 \div 10 = 4$$

Think: How many tens equal 40?

VOCABULARY

dividend
a number that is being divided

divisor
the number by which the dividend is being divided

quotient
the result of a division problem

You should memorize the division rules for 0 and 1.

- 0: Zero divided by any number (except zero) equals zero.
- 1: Any number divided by one equals the same number.

Example 1

Use a model to find 3 ÷ 1.

1. Draw a model.

2. How many groups will there be?   1

3. How many in each group?   3

4. Write the quotient.   3 ÷ 1 = 3

5. Check.   1 × 3 = 3

YOUR TURN!

Use a model to find 6 ÷ 1.

1. Draw a model.

2. How many groups will there be?   _____

3. How many in each group?   _____

4. Write the quotient.   6 ÷ 1 = _____

5. Check.   1 × 6 = _____

Example 2

Find 30 ÷ 10.

1. What number is the divisor? 10

2. How many tens equal 30?   3

3. Write the quotient.   30 ÷ 10 = 3

4. Check.   3 × 10 = 30

YOUR TURN!

Find 70 ÷ 10.

1. What number is the divisor?   _____

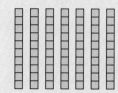

2. How many tens equal 70?   _____

3. Write the quotient.   70 ÷ 10 = _____

4. Check.   7 × 10 = _____

GO ON

Who is Correct?

Find 90 ÷ 10.

Angelina
90 ÷ 10
= 9

James
90 ÷ 10
= 90

Lisa
90 ÷ 10
= 900

Circle correct answer(s). Cross out incorrect answer(s).

 Guided Practice

Use a model to find each quotient.

1 $5 \div 1 =$ _____

How many groups will there be? _____

How many in each group? _____

2 $7 \div 7 =$ _____

How many groups will there be? _____

How many in each group? _____

Step by Step Practice

3 Find $80 \div 10$.

Step 1 What number is the divisor? _____

Step 2 How many tens equal 80? _____

Step 3 Write the quotient. $80 \div 10 =$ _____

Step 4 Check. _____ $\times\ 10 =$ _____

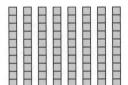

Find each quotient. If the quotient is not possible, write *not possible*.

4 $90 \div 10 = $ _____

Check. _____ × _____ = _____

5 $70 \div 10 = $ _____

6 $10 \div 10 = $ _____

7 $40 \div 10 = $ _____

8 $60 \div 10 = $ _____

9 $9 \div 1 = $ _____

10 $6 \div 1 = $ _____

11 $15 \div 0 = $ _____

12 $3 \div 1 = $ _____

13 $17 \div 1 = $ _____

14 $4 \div 0 = $ _____

15 $0 \div 20 = $ _____

16 $0 \div 2 = $ _____

17 $13 \div 13 = $ _____

18 $8 \div 8 = $ _____

Step *by* Step **Problem-Solving Practice**

Solve.

19 **COINS** There are 10 pennies in each dime. If you have 50 pennies, for how many dimes could you trade?

Understand Read the problem. Write what you know.

There are _____ pennies in a dime.

There are _____ pennies altogether.

Plan Pick a strategy. One strategy is to make a table.

Solve Complete the table.

The table shows that 50 pennies equal

_____ dimes.

Check Does the answer make sense? Look over your solution. Did you answer the question?

Problem-Solving Strategies
- ☐ Draw a model.
- ☐ Use logical reasoning.
- ☑ Make a table.
- ☐ Solve a simpler problem.
- ☐ Work backward.

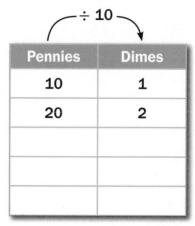

Pennies	Dimes
10	1
20	2

GO ON

20 PACKAGING   There are 10 rolls of paper towels in every jumbo package. If a shopper wants 80 rolls of paper towels, how many packages should he purchase?
Check off each step.

_____ Understand: I underlined key words.

_____ Plan: To solve the problem, I will _____.

_____ Solve: The answer is _____.

_____ Check: I checked my answer by _____.

21 FOOD   The cafeteria used 120 eggs for a recipe. How many dozen eggs did they use?

_____

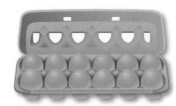
A dozen is 12 eggs.

22 Reflect   Explain why you cannot divide by zero.

_____

_____

_____

 Skills, Concepts, and Problem Solving

Use a model to find each quotient.

23 $10 \div 10 =$ _____

24 $40 \div 10 =$ _____

25 $7 \div 1 =$ _____

26 $50 \div 10 =$ _____

27 $9 \div 9 =$ _____

28 $90 \div 10 =$ _____

Find each quotient. If the quotient is not possible, write *not possible*.

29 $3 \div 0 =$ _____

30 $0 \div 7 =$ _____

31 $1 \div 1 =$ _____

32 $2 \div 1 =$ _____

33 $8 \div 1 =$ _____

34 $4 \div 1 =$ _____

35 $3 \div 3 =$ _____

36 $5 \div 5 =$ _____

37 $60 \div 10 =$ _____

38 $90 \div 1 =$ _____

39 $12 \div 1 =$ _____

40 $15 \div 1 =$ _____

41 $40 \div 0 =$ _____

42 $40 \div 1 =$ _____

43 $30 \div 10 =$ _____

44 $0 \div 80 =$ _____

Solve.

45 **TILES** Marcos is laying tiles on the kitchen floor. The tiles come in boxes of 10. If each tile is 1 square foot, how many boxes will Marcos need to tile a floor that is 80 square feet?

_____

GO ON

Solve.

46 **BASKETBALL** Each storage rack in the school gym can hold 12 basketballs. Mrs. Collins has 12 basketballs. How many storage racks does Mrs. Collins need? What division sentence models her actions?

_____

47 **MODELING** Austin uses base-ten blocks to model a division problem. He takes the one-blocks and groups them into 10 piles of 9 blocks. What division sentence models his actions?

_____

48 **EXERCISE** Joy's grandfather started training on a treadmill machine. By the third month, he trained 10 times longer each day than he did the first month. He trained for 70 minutes a day the third month. How many minutes a day did he train when he began?

_____

49 **RECYCLING** Mauricio has collected 120 aluminum cans. If 10 cans will fit in each blue plastic bag, how many bags will he need to carry all the cans?

_____

EXERCISE Joy's grandfather trains on a treadmill.

Vocabulary Check **Write the vocabulary word that completes each sentence.**

50 In a division problem, the number being divided is the

_____.

51 Zero cannot be a _____ because you cannot divide by zero.

52 The _____ is the result of a division problem.

53 **Writing in Math** Explain why division can be thought of as repeated subtraction.

_____

_____

_____

STOP

Division with 2 through 6

KEY Concept

Division is a process that separates a large group into smaller groups.

You can divide into same-size groups and count the groups. Or, you can divide into a specific number of groups and find the size of each group.

Divide 12 into groups of 3.

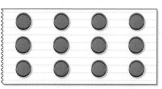

Divide 12 into 3 groups.

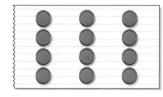

divisor
↓
dividend →12 ÷ 4 = 3← quotient   12 ÷ 3 = 4

VOCABULARY

dividend
a number that is being divided

division
an operation on two numbers in which the first number is split into the same number of equal groups as the second number

divisor
the number by which the dividend is being divided

quotient
the result of a division problem

To solve a division fact, you can use models, a number line, or a related multiplication fact.

Example 1

Use or draw a model to find the quotient of 14 ÷ 2.

1. The dividend is 14, so use 14 counters.

2. The divisor is 2, so make or circle groups of 2.

3. There are 7 groups of 2, so the quotient is 7.

   14 ÷ 2 = 7

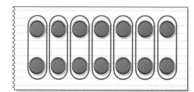

YOUR TURN!

Use or draw a model to find the quotient of 24 ÷ 3.

1. The dividend is _____, so use _____ counters.

2. The divisor is _____, so make or circle groups of _____.

3. There are _____ groups of _____, so the quotient is _____.

   24 ÷ 3 = _____

GO ON

Example 2

Use a number line to find the quotient of 18 ÷ 3.

1. Draw a number line from 0 to 18.
   Draw a point on the number line at 18.

   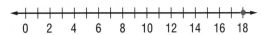

2. Count back 3 and show it with an arrow.
   Draw a point where you land.

   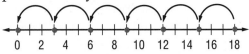

3. Keep counting back 3 and drawing points until you reach 0.

   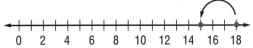

4. Count the jumps. There are 6 jumps, so 18 ÷ 3 = 6.

   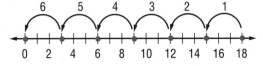

YOUR TURN!

Use a number line to find the quotient of 20 ÷ 5.

1. Draw a number line from 0 to _____.
   Draw a point at _____.

   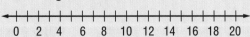

2. Count back 5 and show it with an arrow.
   Draw a point where you land.
   Where will you draw the point? _____

   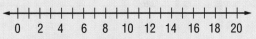

3. Keep counting back and drawing points until you reach _____.

   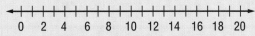

4. Count the jumps. There are _____ jumps, so 20 ÷ 5 = _____.

   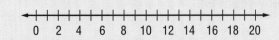

Example 3

Use a related multiplication fact to find the quotient of 24 ÷ 6.

1. Think of a multiplication fact with a factor of 6 and a product of 24.

   6 × _____ = 24

2. What number multiplied by 6 equals 24?
   4

3. Use the missing number to complete the related division fact.

   24 ÷ 6 = 4

YOUR TURN!

Use a related multiplication fact to find the quotient of 21 ÷ 3.

1. Think of a multiplication fact with a factor of 3 and a product of 21.

   _____

2. What number multiplied by 3 equals 21?

   _____

3. Use the missing number to complete the related division fact.

   _____

Who is Correct?

Find the quotient of 28 ÷ 4.

Joe
4 × ___ = 28

4 × 7 = 28

So, 28 ÷ 4 = 7.

Lela
28 ÷ 4 = 12

Rasha
4 × ___ = 28

4 × 8 = 28

So, 28 ÷ 4 = 8.

Circle correct answer(s). Cross out incorrect answer(s).

▶ Guided Practice

Use or draw a model to find each quotient.

1 32 ÷ 4 = _____

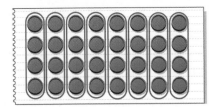

_____ groups of 4

2 16 ÷ 4 = _____

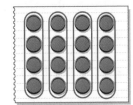

_____ groups of 4

3 27 ÷ 3 = _____

4 18 ÷ 2 = _____

5 20 ÷ 5 = _____

6 24 ÷ 6 = _____

GO ON

Use a number line to find each quotient.

7 $12 \div 3 =$ _____

0 1 2 3 4 5 6 7 8 9 10 11 12

8 $18 \div 6 =$ _____

0 2 4 6 8 10 12 14 16 18

9 $25 \div 5 =$ _____

0 5 10 15 20 25

Step by Step Practice

10 Use a related multiplication fact to find $48 \div 6$.

Step 1 Think of a multiplication fact with a factor of 6 and a product of 48.

$6 \times$ _____ $= 48$

Step 2 Complete the multiplication fact.

$6 \times$ _____ $= 48$

Step 3 Use the missing factor to complete the related division fact.

_____ $\div$ _____ $=$ _____

Use a related multiplication fact to find each quotient.

11 $10 \div 2$

_____ $\times$ _____ $=$ _____

_____ $\div$ _____ $=$ _____

12 $15 \div 5$

_____ $\times$ _____ $=$ _____

_____ $\div$ _____ $=$ _____

13 $40 \div 5$

$5 \times$ _____ $= 40$

$40 \div 5 =$ _____

14 $42 \div 6$

$6 \times$ _____ $= 42$

$42 \div 6 =$ _____

15 $12 \div 3$

$3 \times$ _____ $= 12$

$12 \div 3 =$ _____

16 $36 \div 4$

$4 \times$ _____ $= 36$

$36 \div 4 =$ _____

Step by Step Problem-Solving Practice

Solve.

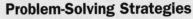

Problem-Solving Strategies

☑ Draw a diagram.
☐ Use logical reasoning.
☐ Make a table.
☐ Solve a simpler problem.
☐ Work backward.

17 **BAKING**   Mighty Muffins makes 36 muffins at a time. They are baked in tins that hold 6 muffins each. How many muffin tins are used to make the muffins?

Understand   Read the problem. Write what you know.

A total of _____ muffins are made at one time.

The muffins are made in tins that hold _____ muffins.

Plan   Pick a strategy. One strategy is to draw a diagram.

Solve   Use 36 counters or draw 36 dots to represent the muffins. Circle groups of 6 to show the muffin tins.

There are _____ muffin tins.

Check   Use a related multiplication fact to solve.

$6 \times$ ____ $= 36$, so ____ $\div$ ____ $=$ ____.

18 **MONEY**   Bret has 35 pennies. He wants to change them for nickels. How many nickels can Bret get? (A nickel is worth 5 pennies.)
Check off each step.

_____ Understand: I underlined key words.

_____ Plan: To solve the problem, I will _____.

_____ Solve: The answer is _____.

_____ Check: I checked my answer by using _____.

GO ON

19 STUDENTS   Mr. Harris wants to divide his 28 students into equal groups of 4 students each. How many groups of 4 can he make? _____

20 ART   The art museum has 72 pieces to place evenly into 6 exhibit halls. How many pieces will go into each exhibit hall? _____

21 Reflect   Explain and show how you would use a model and number line to find the quotient for 18 ÷ 3.

_____

_____

▶ Skills, Concepts, and Problem Solving

Use or draw a model to find each quotient.

22 35 ÷ 5 = _____

23 20 ÷ 2 = _____

24 20 ÷ 4 = _____

25 21 ÷ 3 = _____

Use a number line to find each quotient.

26 10 ÷ 5 = _____

```
0            5          10
```

27 15 ÷ 3 = _____

```
0   3   6   9   12  15
```

28 24 ÷ 6 = _____

```
0    6    12   18   24
```

Use a related multiplication fact to find each quotient.

29 40 ÷ 4 = _____

30 22 ÷ 2 = _____

31 54 ÷ 6 = _____

32 60 ÷ 5 = _____

33 FOOD DRIVE   Alma's scout troop collects 45 cans for the food drive. If the 5 scouts collect the same number of cans, how many cans does each scout collect?

_____

34 PARTY PLANNING   Ms. Lane is planning to have a celebration for 72 students. If six people sit at each table, how many tables will Ms. Lane need?

_____

Vocabulary Check   **Write the vocabulary word that completes each sentence.**

35 The result of a division problem is called a _____.

36 _____ is an operation on two numbers in which the first number is split into the same number of equal groups as the second number.

37 Writing in Math   What division problem can you solve using the related multiplication fact of 6 × 10 = 60? Tell how you know.

_____

_____

_____

 Spiral Review

Solve.   (Lesson 6-1, p. 234)

38 FOOD   The chef wants to make sure there is one piece of pizza for every student. If he cuts each pizza into 10 slices, how many whole pizzas will he need to have enough for 120 students? _____

39 GRAPHING   While making a scale for her graph, Ebony wants each square to represent 10 years. If she plots a point that represents 70 years, how many squares high will it be? _____

Progress Check 1 (Lessons 6-1 and 6-2)

Find each quotient. If the quotient is not possible, write *not possible*.

1   $0 \div 6 =$ _____

2   $25 \div 25 =$ _____

3   $60 \div 10 =$ _____

4   $70 \div 1 =$ _____

5   $15 \div 0 =$ _____

6   $8 \div 8 =$ _____

7   $42 \div 1 =$ _____

8   $24 \div 0 =$ _____

9   $33 \div 33 =$ _____

10   $50 \div 1 =$ _____

Use a model to find each quotient.

11   $18 \div 3 =$ _____

12   $16 \div 2 =$ _____

Use a number line to find each quotient.

13   $24 \div 6 =$ _____

0   6   12   18   24

14   $15 \div 5 =$ _____

Solve.

15   **GARDENING**   Brian is planting 35 flowers. He wants to have 5 equal rows of flowers. How many flowers will he put in each row?

_____

16   **PARTY**   Liliana placed 30 items into 10 favor bags. She placed the same number of items in each bag. How many items did she place in each bag?

_____

Division with 7 through 12

KEY Concept

There are several phrases that represent **division**.

$48 \div 8$ can mean:

- the **quotient** of 48 and 8
- 48 divided by 8
- a number multiplied by 8 is 48

Division is the **inverse operation** of multiplication.

You can use multiplication to help you find or remember a division fact.

$8 \times 6 = 48$, so $48 \div 8 = 6$.

VOCABULARY

dividend
  a number that is being divided

division
  an operation on two numbers in which the first number is split into the same number of equal groups as the second number

divisor
  the number by which the dividend is being divided

inverse operations
  operations which undo each other

quotient
  the result of a division problem

You have used arrays to solve multiplication facts. Arrays can help you solve division facts, too.

Example 1

Draw an array to find the quotient of 28 ÷ 7.

1. Draw an array with 28 rectangles in 7 rows.

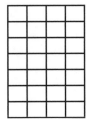

2. Count the number of columns. There are 4 columns, so the quotient is 4.

3. Check by multiplying the quotient by the divisor. The product should be the same as the dividend.

   $4 \times 7 = 28$, so $28 \div 7 = 4$.

YOUR TURN!

Draw an array to find the quotient of 40 ÷ 8.

1. Draw an array with _____ rectangles in _____ rows.

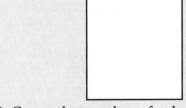

2. Count the number of columns. There are _____ columns. The quotient is _____.

3. Check by multiplying the quotient by the divisor. The product should be the same as the dividend.

   _____ × _____ = _____

GO ON

Example 2

Use a related multiplication fact to find the quotient of 60 ÷ 10.

1. Think of a multiplication fact with a factor of 10 and a product of 60.

   10 × _____ = 60

2. What number multiplied by 10 equals 60? **6**

3. Use the missing number to complete the related division fact.   60 ÷ 10 = 6

YOUR TURN!

Use a related multiplication fact to find the quotient of 72 ÷ 9.

1. What multiplication fact has a factor of 9 and a product of 72? _____

2. What number multiplied by 9 equals 72? _____

3. Use the missing number to complete the related division fact. _____

Example 3

Use or draw models to find the quotient of 24 ÷ 12.

1. The dividend is 24, so use 24 counters.

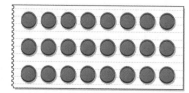

2. The divisor is 12, so make or circle groups of 12.

3. There are 2 groups of 12, so the quotient is 2.

   24 ÷ 12 = 2

YOUR TURN!

Use or draw models to find the quotient of 32 ÷ 8.

1. The dividend is _____, so use _____ counters.

2. The divisor is _____, so make or circle groups of _____.

3. There are _____ groups of _____, so the quotient is _____.

   32 ÷ 8 = _____

Who is Correct?

Find the quotient of 21 ÷ 3.

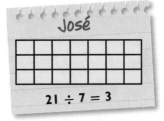

Boris

21 ÷ 3 = 6

Meliah

7 × ___ = 21
7 × 3 = 21
So, 21 ÷ 7 = 3

José

21 ÷ 7 = 3

Circle correct answer(s). Cross out incorrect answer(s).

▶ Guided Practice

Draw an array to find each quotient.

1 42 ÷ 7 = _____

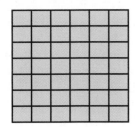

2 54 ÷ 9 = _____

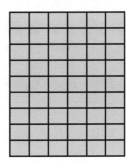

3 44 ÷ 11 = _____

4 27 ÷ 9 = _____

Use a related multiplication fact to find each quotient.

5 108 ÷ 12

12 × _____ = 108

108 ÷ 12 = _____

6 49 ÷ 7

7 × _____ = 49

49 ÷ 7 = _____

GO ON

Use a related multiplication fact to find each quotient.

7 $56 \div 8$

$8 \times$ _____ $= 56$

$56 \div 8 =$ _____

8 $132 \div 11$

$11 \times$ _____ $= 132$

$132 \div 11 =$ _____

Step by Step Practice

9 Use or draw a model to find $60 \div 12$.

Step 1 Draw models to represent the dividend.

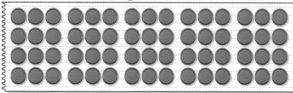

Step 2 Circle groups to represent the divisor. The divisor is 12, so circle groups of 12.

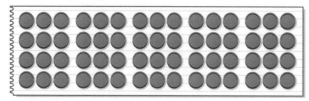

Step 3 Count the groups. There are _____ groups.

$60 \div 12 =$ _____

Use or draw models to find each quotient.

10 $45 \div 9 =$ _____

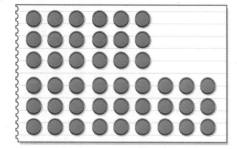

11 $36 \div 12 =$ _____

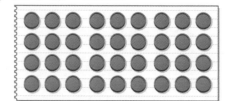

12 $56 \div 7 =$ _____

13 $90 \div 9 =$ _____

Step by Step Problem-Solving Practice

Copyright © Glencoe/McGraw-Hill, a division of The McGraw-Hill Companies, Inc.

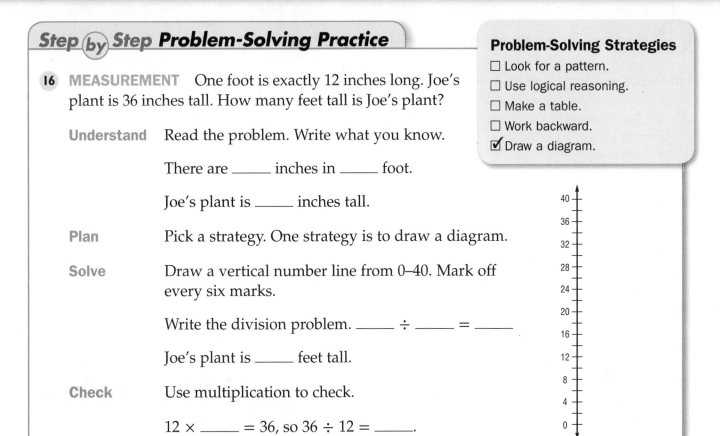

Problem-Solving Strategies
☐ Look for a pattern.
☐ Use logical reasoning.
☐ Make a table.
☐ Work backward.
☑ Draw a diagram.

16 **MEASUREMENT**   One foot is exactly 12 inches long. Joe's plant is 36 inches tall. How many feet tall is Joe's plant?

Understand   Read the problem. Write what you know.

There are _____ inches in _____ foot.

Joe's plant is _____ inches tall.

Plan   Pick a strategy. One strategy is to draw a diagram.

Solve   Draw a vertical number line from 0–40. Mark off every six marks.

Write the division problem. _____ ÷ _____ = _____

Joe's plant is _____ feet tall.

Check   Use multiplication to check.

$12 \times$ _____ = 36, so $36 \div 12 =$ _____.

17 **SOCCER**   A group of 84 soccer players will be separated into equal teams of 12. How many teams will there be? Check off each step.

_____ Understand: I underlined key words.

_____ Plan: To solve the problem, I will _____.

_____ Solve: The answer is _____.

_____ Check: I checked my answer by using _____.

18 **SWIMMING**   There are 81 students in the swim class. The game they are playing needs teams with 9 students on each team. How many teams will there be?

_____

19 **Reflect**   Explain how to find the quotient using an array.

_____

GO ON

 **Skills, Concepts, and Problem Solving**

Find each quotient. Use an array or a model.

20 $14 \div 7 =$ _____

21 $45 \div 9 =$ _____

Use a related multiplication fact to find each quotient.

22 $48 \div 12 =$ _____

23 $72 \div 8 =$ _____

24 $99 \div 11 =$ _____

25 $63 \div 9 =$ _____

26 GRAPHING   A pictograph uses one book to represent 10 books collected for a book drive. How many symbols are needed to represent the 80 books that Joann collected?

_____

27 SHOPPING   How many $12 CDs can Peni buy

for $84? _____

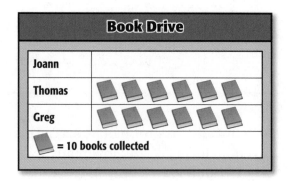

Book Drive

Joann	
Thomas	📚📚📚📚📚📚
Greg	📚📚📚📚📚

📖 = 10 books collected

Vocabulary Check   **Write the vocabulary word that completes each sentence.**

28 _____ are operations which undo each other.

29 A(n) _____ is a number being divided.

30 Writing in Math   Write a real-world problem that uses the division fact $72 \div 8$.

_____

_____

 Spiral Review

Find each quotient. (Lesson 6-2, p. 241)

31 $24 \div 3 =$ _____

32 $42 \div 6 =$ _____

33 $36 \div 4 =$ _____

KEY Concept

Sometimes **division** can produce an outcome of a **quotient** with a remainder. A **remainder** is an amount left over after dividing into equal groups. The remainder can be treated in different ways.

Ignore the remainder.
Alberto has $7. How many $3 markers can he buy?

Alberto can buy 2 markers.

Figure in as fractional parts.
There are 7 apples to be shared among 3 people. How many apples does each person get?

Each person gets $2\frac{1}{3}$ apples.

Add 1 to the quotient.
A roller coaster car holds 3 people. How many cars are needed for 7 people?

3 cars are needed.

VOCABULARY

division
an operation on two numbers in which the first number is split into the same number of equal groups as the second number

fraction
a number that represents part of a whole or part of a set

quotient
the result of a division problem

remainder
the number that is left after one whole number is divided by another

GO ON

Example 1

Use models to find 14 ÷ 4. Identify the remainder.

1. Use counters to represent the dividend. How many counters will you use? **14**

2. Arrange the counters into the number of groups indicated by the divisor. How many groups can you make? **4**

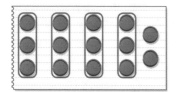

3. How many counters in each group? **3**
   Did you use all the counters? **No**
   How many counters remain? **2**

4. The quotient of 14 ÷ 4 is 3 with a remainder of 2.
   14 ÷ 4 = 3 R2

YOUR TURN!

Use models to find 19 ÷ 3. Identify the remainder.

1. Use counters to represent the dividend.

   How many counters did you use? _____

2. Arrange the counters into the number of groups indicated by the divisor. How many groups can you make? _____

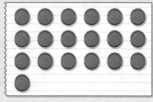

3. How many counters in each group? _____

   Did you use all the counters? _____

   How many counters remain? _____

4. What is the quotient of 19 ÷ 3?

   _____ R _____

Example 2

Draw a picture to find 23 ÷ 5. Show the remainder.

1. Draw a picture to represent the dividend.

2. The divisor is 5. Circle groups of 5.

3. How many circles? **4**

4. How many marks are not inside a circle? **3**

5. The quotient of 23 ÷ 5 is 4 with a remainder of 3.
   23 ÷ 5 = 4 R3

YOUR TURN!

Draw a picture to find 13 ÷ 2. Show the remainder.

1. Draw a picture to represent the dividend.

2. The divisor is _____. Circle groups of _____.

3. How many circles? _____.

4. How many marks are not inside a circle? _____

5. The quotient of 13 ÷ 2 = _____.

Who is Correct?

Find 25 ÷ 6. Identify the remainder.

Benjamin
25 ÷ 6
= 4 R1

Hiranya
25 ÷ 6
= 4

Mykia
25 ÷ 6
= 5

Circle correct answer(s). Cross out incorrect answer(s).

▶ Guided Practice

Use models to find the quotient. Show the remainder.

1 31 ÷ 7 = _____

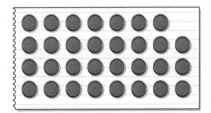

There are _____ groups of

_____ counters and _____
counters not in a group.

2 42 ÷ 5 = _____

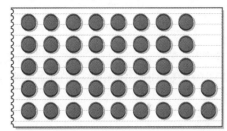

There are _____ groups of

_____ counters and _____
counters not in a group.

3 35 ÷ 8 = _____

4 21 ÷ 9 = _____

5 53 ÷ 7 = _____

6 46 ÷ 6 = _____

GO ON

7 Draw a picture to find the quotient of 43 ÷ 5. Show the remainder.

Step 1 Draw models to represent the dividend.

Step 2 Circle groups to represent the divisor.
The divisor is 5, so circle groups of 5.

Step 3 How many circles? _____

Step 4 There are _____ marks
not in a group, so the remainder is _____.

43 ÷ 5 = _____

Draw a picture to find the quotient. Show the remainder.

8 17 ÷ 8 = _____

9 26 ÷ 8 = _____

10 30 ÷ 3 = _____

11 34 ÷ 6 = _____

Step by Step Problem-Solving Practice

Problem-Solving Strategies

☑ Draw a picture.
☐ Look for a pattern.
☐ Guess and check.
☐ Act it out.
☐ Solve a simpler problem.

12 CRAFTS Ebony is making key chains for her friends. She needs 6 beads for each keychain. She has 22 beads. How many key chains can Ebony make?

Understand Read the problem. Write what you know.

Ebony has _____ beads in all.

She needs _____ beads for each key chain.

Plan Pick a strategy. One strategy is to draw a picture.

Solve Draw 22 beads. Each key chain needs _____

beads, so circle groups of _____.

Count the circled groups. There are _____ groups,

with _____ beads remaining.

Are there enough beads remaining to make another

key chain? _____

Ebony can make _____ key chains.

Check Use multiplication and addition to check.

$$\underline{\hspace{2cm}} \times \underline{\hspace{2cm}} = \underline{\hspace{2cm}}$$
number of   number of beads   number of beads
keychains   on each chain   in keychains

$$\underline{\hspace{2cm}} + \underline{\hspace{2cm}} = \underline{\hspace{2cm}}$$
number of   remaining   beads
beads   beads

If the total matches the original dividend, the division is correct.

GO ON

13 **PACKING** Uma has 37 books. She can fit 8 books into each carton. How many cartons does Uma need to pack all of the books? Check off each step.

_____ Understand: I underlined key words.

_____ Plan: To solve the problem, I will _____.

_____ Solve: The answer is _____.

_____.

_____ Check: I checked my answer by using _____.

14 **POSTCARDS** The Science and Industry Museum sells postcards for $3. Stephen has $29 to spend. If he buys as many postcards as he can, how much money will he have left? _____

15 **Reflect** Explain why the remainder can be ignored when using money.

_____

_____

 Skills, Concepts, and Problem Solving

Use models to find each quotient. Identify the remainder.

16 11 ÷ 5 = _____

17 29 ÷ 8 = _____

Draw pictures to find each quotient. Show the remainder.

18 15 ÷ 2 = _____

19 39 ÷ 6 = _____

20 28 ÷ 11 = _____

21 43 ÷ 9 = _____

Solve. Explain what to do with the remainder.

22 **SHOPPING** How many $17 sweatshirts can Serena buy with $53?

_____

_____

23 **FLOWER BOXES** Ruben buys 42 flowers for his garden. He wants to put 5 flowers in each flower box. How many flower boxes does he need to plant all of the flowers?

_____

_____

24 **DINOSAURS** An artist makes models of dinosaurs. He has 59 dinosaurs. He packages them in boxes of 4. How many boxes can the artist fill?

_____

_____

Vocabulary Check **Write the vocabulary word that completes each sentence.**

25 A(n) _____ is the number that is left after one whole number is divided by another.

26 A(n) _____ is the result of a division problem.

27 **Writing in Math** When buying items, you can (always, sometimes, never) use the remaining money to buy an additional item. Explain your reasoning.

_____

_____

▶ **Spiral Review**

Find each quotient. (Lesson 6-2, p. 241 and 6-3, p. 249)

28 $36 \div 4 =$ _____

29 $18 \div 2 =$ _____

30 $24 \div 3 =$ _____

31 $64 \div 8 =$ _____

32 $54 \div 9 =$ _____

33 $40 \div 5 =$ _____ STOP

Use or draw an array to find each quotient.

1   $32 \div 8 =$ _____

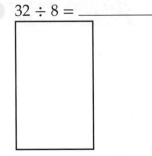

2   $42 \div 7 =$ _____

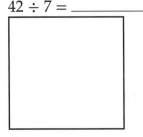

Use a related multiplication fact to find each quotient.

3   $36 \div 4 =$ _____

_____

4   $50 \div 5 =$ _____

_____

5   $33 \div 3 =$ _____

_____

6   $48 \div 6 =$ _____

_____

Use models to find each quotient. Identify the remainder.

7   $17 \div 2 =$ _____

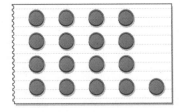

8   $26 \div 4 =$ _____

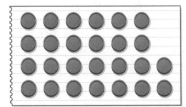

Solve.

9   **PARTY**   Tommy buys flags at the store. Each flag cost $6. How many flags can he buy for $90?

_____

10 **MUSEUM**   A museum has 203 wildlife photographs. Anessa wants to display them in a groups of 8. How many groups of 8 can she make?

_____

Long Division with Single-Digit Divisors

KEY Concept

Before you start a long division problem, you should estimate so that you can check your quotient for reasonableness. To estimate, use the division facts for 1 through 10.

$$\begin{array}{r} 1,000 \\ 4\overline{)4,000} \end{array}$$

Round the dividend to the nearest 1,000 that is divisible by 4.

$$\begin{array}{r} 1,054 \\ 4\overline{)4,216} \\ \underline{-4} \\ 02 \\ \underline{-0} \\ 21 \\ \underline{-20} \\ 16 \\ \underline{-16} \\ 0 \end{array}$$

1. Divide: 4 ÷ 4
2. Multiply: 4 × 1
3. Subtract: 4 − 4
4. Bring down the 2.
5. Repeat Steps 1–4 until there is nothing to bring down.

VOCABULARY

dividend
the number that is being divided

divisor
the number by which the dividend is being divided

quotient
the result of a division problem

Estimate before you divide and compare your quotient to the estimate.

Example 1

Find 1,896 ÷ 2. Estimate first.

1. Estimate. **1,800 ÷ 3 = 600**

2. Rewrite the problem in vertical format.

$$3\overline{)1,896}$$

3. Look at the first digit. Since the divisor is greater than the first digit, look at the first two digits. What number multiplied by 3 is 18? **6** Multiply. Subtract.

$$\begin{array}{r} 6 \\ 3\overline{)1,896} \\ \underline{-18} \\ 0 \end{array}$$

4. What number multiplied by 3 is 9? **3** Multiply. Subtract.

$$\begin{array}{r} 63 \\ 3\overline{)1,896} \\ \underline{-18} \\ 09 \\ \underline{-9} \\ 0 \end{array}$$

5. What number multiplied by 3 is 6? **2** Multiply. Subtract.

$$\begin{array}{r} 632 \\ 3\overline{)1,896} \\ \underline{-18} \\ 09 \\ \underline{-9} \\ 06 \\ \underline{-6} \\ 0 \end{array}$$

6. The quotient is 632. Compare to the estimate for reasonableness.

$$\begin{array}{r} 632 \\ 3\overline{)1,896} \end{array}$$

GO ON

YOUR TURN!

Find 4,025 ÷ 5.

1. Estimate. 4,000 ÷ 5 = _____

2. Rewrite the problem in vertical format.

3. Look at the first digit. Since the divisor is greater than the first digit, look at the first two digits. What number multiplied by 5 is 40? _____
   Multiply. Subtract.

4. Multiply. Subtract. Bring down the next number in the dividend.

5. Multiply. Subtract.

6. The quotient is _____.
   Compare to the estimate for reasonableness.

$$\begin{array}{r} \\ \overline{)} \\ \overline{} \\ \overline{} \\ \overline{} \end{array}$$

Who is Correct?

Find 125 ÷ 6. Show the remainder.

Miles

120 ÷ 6 = 20

$$\begin{array}{r} 20 \\ 6)\overline{125} \\ \underline{12} \\ 05 \end{array}$$

20 R5

Simone

$$\begin{array}{r} 20 \\ 6)\overline{125} \\ \underline{12} \\ 05 \end{array}$$

6 − 5 = 1

20 R1

Rebeca

$$\begin{array}{r} 112 \\ 6)\overline{125} \\ \underline{6} \\ 65 \\ \underline{60} \\ 5 \end{array}$$

112 R5

Circle correct answer(s). Cross out incorrect answer(s).

 Guided Practice

Estimate each quotient. Then divide and show the remainder.

1 316 ÷ 8

_____

2 182 ÷ 3

_____

3 8,583 ÷ 6

_____

4 5,486 ÷ 5

_____

Step by Step Practice

Find each quotient. Estimate first.

5 6,082 ÷ 2

Step 1 Estimate. _____

Step 2 Rewrite the problem in vertical format.

Step 3 What number multiplied by 2 is 6? _____
Multiply. Subtract.

Step 4 Since 0 < 3, there is not enough to divide. So, put 0 in
the hundreds place. Bring down the next number.

Step 5 What number multiplied by 2 is 8? _____
Multiply. Subtract.

Step 6 What number multiplied by 2 is 2? _____
Multiply. Subtract.

Step 7 The quotient is _____.
Compare to the estimate for reasonableness.

Find each quotient. Estimate first. Show the remainder if there is one.

6 1,131 ÷ 3

_____

7 1,214 ÷ 2

_____

8 1,210 ÷ 5

_____

9 1,080 ÷ 4

_____

10 483 ÷ 7

_____

11 846 ÷ 6

_____

12 280 ÷ 9

_____

13 994 ÷ 8

_____

14 615 ÷ 12

_____

15 478 ÷ 11

_____

GO ON

Step by Step Problem-Solving Practice

Solve.

Problem-Solving Strategies
☐ Draw a diagram.
☑ Use logical reasoning.
☐ Guess and Check.
☐ Solve a simpler problem.
☐ Work backward.

16 PACKAGING   The principal is buying a ribbon for all of the 296 honor roll students. The ribbons come with 8 in each box. How many boxes will be needed to buy?

Understand   Read the problem. Write what you know.

There are _____ honor roll students.

Ribbons are sold by _____ per box.

Plan   Pick a strategy. One strategy is to use logical reasoning.

Solve   Estimate the given values.
There are about 10 ribbons in each box.
There are about 300 students on the honor roll.

$300 \div 10 =$ _____

For every box the principal buys, there will

be enough ribbons for _____ students.

This is a problem about equal grouping, which means it is a division problem.

Divide the total needed by the number of pins

in a box. _____

The principal needs to buy _____ boxes of ribbons.

Check   Use the estimated values to check if the answer is reasonable.

17 SNACKS   Gail made 138 big cookies. She can get 6 cookies in one box. How many boxes will she need to package all cookies? Check off each step.

_____ Understand: I underlined key words.

_____ Plan: To solve the problem, I will _____.

_____ Solve: The answer is _____.

_____ Check: I checked my answer by _____.

18 **GRADUATION**   There will be 9,128 friends and relatives attending a graduation ceremony this weekend. If each graduate has 4 people attending the ceremony, how many students are graduating?

_____

19 **Reflect**   Division is a method of repeated subtraction. Explain this sentence and illustrate it with an example.

_____

_____

▶ Skills, Concepts, and Problem Solving

Find each quotient. Show the remainder.

20 $219 \div 4 =$ _____

21 $445 \div 7 =$ _____

22 $1,655 \div 2 =$ _____

23 $3,140 \div 9 =$ _____

Find each quotient. Estimate first. Show the remainder if there is one.

24 $3,036 \div 3$

25 $1,768 \div 2$

26 $990 \div 5$

27 $936 \div 4$

28 $1,672 \div 8$

29 $1,896 \div 6$

30 $3,578 \div 9$

31 $2,626 \div 8$

GO ON

Solve.

GAMES   A game that 100 students are playing needs teams with 12 students on each team.

32 How many complete teams will there be?

_____

33 How many more students would be needed to make another team?

_____

Vocabulary Check   **Write the vocabulary word that completes each sentence.**

34 A remainder must be less than the _____.

35 When the division problem is written as a fraction, the

_____ is the numerator.

36 **Writing in Math**   Explain the first step in long division when the divisor is greater than the first digit of the dividend.

_____

_____

▶ Spiral Review

Use models to find each quotient. Show the remainder.

37 $50 \div 5 =$ _____

38 $32 \div 8 =$ _____

39 $34 \div 4 =$ _____

40 $22 \div 6 =$ _____

STOP

Long Division with Two-Digit Divisors

KEY Concept

The steps to dividing with a two-digit divisor are the same as dividing by a single-digit divisor.

Before solving a division problem, you can use rounding and estimation to predict your answer. After solving a division problem, you can check your answer using estimation.

Estimate: $900 \div 30 = 30$

$$
\begin{array}{r}
32 \\
28\overline{)896} \\
-84 \\
\hline
56 \\
-56 \\
\hline
0
\end{array}
$$

1. Divide: $89 \div 28$
2. Multiply: 28×3
3. Subtract: $89 - 84$
4. Bring down the 6.

  1. Divide: $56 \div 28$
  2. Multiply: 28×2
  3. Subtract: $56 - 56$
  4. Nothing to bring down.

VOCABULARY

dividend
a number that is being divided

division
an operation on two numbers in which the first number is split into the same number of equal groups as the second number

divisor
the number by which the dividend is being divided

quotient
the result of a division problem

Division and multiplication facts can help you predict an estimate for the quotient.

Example 1

Find 228 ÷ 22. Estimate first.

1. Estimate. $200 \div 20 = 10$

2. Rewrite the problem in vertical format.

3. Look at the first two digits in the dividend.

   What number multiplied by 22 is 22? **1**

   Multiply. Subtract.

4. What number multiplied by 22 is 8? **0**

5. What is the remainder? **8**

6. The quotient is **10** with a remainder of **8**.

$$
\begin{array}{r}
10 \text{ R8} \\
22\overline{)228} \\
-22 \\
\hline
08 \\
-00 \\
\hline
8
\end{array}
$$

GO ON

YOUR TURN!

Find 784 ÷ 18.

1. Estimate. 800 ÷ 20 = _____

2. Rewrite the problem in vertical format.

3. Look at the first two digits in the dividend.

$\overline{)}$

   What number multiplied by 18 is close to 78? _____

   $\underline{}$

   Multiply. Subtract.

   $\underline{}$

4. What number multiplied by 18 is close to 64? _____

5. What is the remainder? _____

6. The quotient is _____ with a remainder of _____.

Who is Correct?

Find 897 ÷ 89.

Tiffany

```
    1 R7
89)897
  - 89
    07
  - 00
    7
```

Asad

```
    1 R7
89)897
  - 89
```

Emilio

```
   10 R7
89)897
  - 89
    07
  - 00
    7
```

Circle correct answer(s). Cross out incorrect answer(s).

 Guided Practice

Divide. Use your estimate to check each quotient.

1 400 ÷ 54

   Estimate: _____

   Quotient: _____

2 823 ÷ 91

   Estimate: _____

   Quotient: _____

3 1,020 ÷ 49

   Estimate: _____

   Quotient: _____

4 4,295 ÷ 58

   Estimate: _____

   Quotient: _____

5   Find the quotient of 642 ÷ 21. Use multiplication and addition
    to check your answer.

Step 1   Estimate. _____

Step 2   Rewrite the problem in vertical format.

Step 3   Look at the first two digits in the dividend.

Step 4   Multiply. Subtract. Bring down the next number in
             the dividend.

Step 5   Multiply. Subtract.

Step 6   The quotient is _____ with a remainder of _____.

Find each quotient. Estimate first. Show the remainder if there is one.

6   375 ÷ 46

_____

7   289 ÷ 31

_____

8   751 ÷ 30

_____

9   1,576 ÷ 42

_____

10   539 ÷ 47

_____

11   388 ÷ 39

_____

12   416 ÷ 22

_____

13   649 ÷ 84

_____

GO ON

14 WORK HOURS   Trey worked 1,800 hours in 2009, and worked the same number of hours for 50 weeks. How many hours per week did Trey work?

Understand   Read the problem. Write what you know.

Trey worked _____ hours in 2009.

He worked _____ weeks.

Plan   Pick a strategy. One strategy is to use estimation.

Solve   Estimate. 1,800 is about _____.

_____ ÷ 50 is _____, so the quotient should be

about _____.

Divide to find the actual quotient.

1,800 ÷ 50 = 36, so Trey worked 36 hours per week.

Check   Use multiplication to check your answer.

15 CLASS SIZE   There are 672 students taking Algebra I at Rockville Middle School. There are 28 students in each class. How many classes are there?
Check off each step.

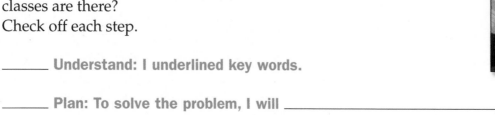

_____ **Understand: I underlined key words.**

_____ **Plan: To solve the problem, I will** _____.

_____ **Solve: The answer is** _____.

_____ **Check: I checked my answer by** _____.

16 **SERVICE** Talia helped serve holiday meals to 47 families. If she served 255 pounds of food, how many pounds of food did each family receive?

_____

17 **Reflect** How is dividing by two-digit divisors similar to dividing by single-digit divisors?

_____

_____

▶ Skills, Concepts, and Problem Solving

Divide. Estimate first. Show the remainder if there is one.

18 $823 \div 20$

_____

19 $523 \div 13$

_____

20 $940 \div 27$

_____

21 $782 \div 38$

_____

22 $4,836 \div 24$

_____

23 $3,105 \div 58$

_____

Find each quotient. Show the remainder.

24 $561 \div 78 =$ _____

25 $855 \div 42 =$ _____

26 $183 \div 15 =$ _____

27 $987 \div 34 =$ _____

GO ON

28 **PACKAGING** Midland School receives 864 textbooks in 36 cartons. How many textbooks are in each carton?

_____

29 **MUSIC** Drew downloads 714 songs on CDs. Each CD holds 21 songs. How many CDs does Drew need to download every song?

_____

Vocabulary Check **Write the vocabulary word that completes each sentence.**

30 The number by which the dividend is being divided is the _____.

31 **Writing in Math** How is dividing by one-digit numbers similar to dividing by two-digit numbers? How is it different?

_____

_____

_____

 Spiral Review

Find each quotient. (Lessons 6-2, p. 241; 6-3, p. 249; 6-4, p. 255)

32 $55 \div 5 =$ _____

33 $60 \div 4 =$ _____

34 $36 \div 6 =$ _____

35 $72 \div 12 =$ _____

36 $114 \div 11 =$ _____

37 $88 \div 9 =$ _____

38 $324 \div 6 =$ _____

39 $75 \div 4 =$ _____

40 $400 \div 8 =$ _____

STOP

Find each quotient.

1. 8)458

2. 7)217

3. 4)356

4. 9)1,246

5. 12)420

6. 11)209

7. 12)264

8. 11)374

Find each quotient. Estimate first. Show the remainder if there is one.

9. 4,617 ÷ 9

_____

10. 1,416 ÷ 13

_____

11. 6,096 ÷ 28

_____

12. 3,949 ÷ 11

_____

13. 2,199 ÷ 42

_____

14. 4,678 ÷ 8

_____

Solve.

15. **MONEY** Dante has $3,210 in rolls of dimes. Each roll of dimes is worth $5. How many rolls of dimes does he have?

_____

16. **HOMEWORK** During the school year, Ms. Larson has homework checks that are worth 8 points each. Justina received the full 8 points for all of her homework checks. She has a total of 978 homework-check points. How many homework checks has she received? How many extra points does she have?

_____

Vocabulary and Concept Check

dividend, *p. 234*

division, *p. 234*

divisor, *p. 234*

fraction, *p. 255*

inverse operations, *p. 249*

remainder, *p. 255*

Write the vocabulary word that completes each sentence.

1 _____ are opposite operations, which means the operations undo each other.

2 A number that is left after one whole number is divided by another is a(n) _____.

3 The number that follows the division sign in a division sentence is the _____.

4 A number that represents part of a whole and part of a set is a(n) _____.

Label each diagram below. Write the correct vocabulary term in each blank.

5 _____

6 _____

$$34 \div 4 = 9$$

Lesson Review

6-1 Divide by 0, 1, and 10 (pp. 234–240)

Find each quotient. If the quotient is not possible, write *not possible*.

7 $10 \div 10 =$ _____

8 $18 \div 1 =$ _____

9 $65 \div 0 =$ _____

10 $43 \div 1 =$ _____

11 $12 \div 0 =$ _____

12 $80 \div 80 =$ _____

> **Example 1**
>
> **Use a model to find $6 \div 1$.**
>
> 1. Draw a model.
>
>
>
> 2. How many groups will there be? **1**
>
> 3. How many in each group? **6**
>
> 4. Write the quotient. $6 \div 1$
>
> 5. Check. $1 \times 6 = 6$

Find each quotient. If the quotient is not possible, write *not possible*.

13 $2 \div 1 = $ _____

14 $3 \div 0 = $ _____

15 $15 \div 15 = $ _____

16 $85 \div 1 = $ _____

17 $16 \div 0 = $ _____

18 $63 \div 63 = $ _____

6-2 Division with 2 through 6 (pp. 241–247)

Find each quotient.

19 $45 \div 5 = $ _____

20 $38 \div 2 = $ _____

21 $32 \div 4 = $ _____

22 $36 \div 6 = $ _____

23 $27 \div 3 = $ _____

24 $44 \div 4 = $ _____

Example 2

Find $40 \div 10$.

1. What number is the divisor? **10**

2. How many tens equal 40? **4**

3. Write the quotient. $40 \div 10 = 4$

4. Check. $4 \times 10 = 40$

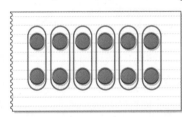
Example 3

Use or draw models to find the quotient of $12 \div 2$.

1. The dividend is 12, so use 12 counters.

2. The divisor is 2, so make or circle groups of 2.

3. There are 6 groups of 2, so the quotient is 6.
   $12 \div 2 = 6$

6-3 Division with 7 through 12 (pp. 249–254)

Find each quotient.

25 $63 \div 7 =$ _____

26 $55 \div 11 =$ _____

27 $32 \div 8 =$ _____

28 $96 \div 12 =$ _____

29 $54 \div 9 =$ _____

30 $49 \div 7 =$ _____

6-4 Remainders (pp. 255–261)

Find each quotient.

31 $19 \div 3 =$ _____

32 $35 \div 4 =$ _____

33 $51 \div 6 =$ _____

34 $17 \div 5 =$ _____

35 $52 \div 7 =$ _____

36 $49 \div 9 =$ _____

Example 4

Draw an array to find the quotient of $24 \div 8$.

1. Draw an array with 24 rectangles in 8 rows.

2. Count the number of columns. There are 3 columns, so the quotient is 3.

3. Check by multiplying the quotient by the divisor. The product should be the same as the dividend.

   $3 \times 8 = 24$, so $24 \div 8 = 3$.

Example 5

Use models to find $20 \div 6$. Identify the remainder.

1. Use counters to represent the dividend. How many counters will you use? **20**

2. Arrange the counters into the number of groups indicated by the divisor. How many groups will you make? **6**

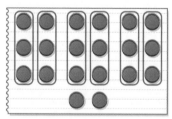

3. How many counters in each group? **3**
   Did you use all the counters? **No**
   How many counters remain? **2**

4. The quotient of $20 \div 6$ is 3 with a remainder of 2, or 3 R2.

6-5 Long Division with Single-Digit Divisors (pp. 263–268)

Find each quotient. Estimate first. Show the remainder.

37 $778 \div 9 =$

_____

38 $1,668 \div 7 =$

_____

39 $1,891 \div 4 =$

_____

40 $557 \div 6 =$

_____

Example 6

Find $6,982 \div 7$. Show the remainder.

1. Estimate. $7,000 \div 7 = 1,000$

2. Rewrite the problem in vertical format.

3. Look at the first two digits. What number multiplied by 7 is close to 69? $7 \times 9 = 63$ Multiply. Subtract.

4. What number multiplied by 7 is close to 68?
$7 \times 9 = 63$
Multiply. Subtract.

5. What number multiplied by 7 is close to 52?
$7 \times 7 = 49$
Multiply. Subtract.

$$\begin{array}{r} 997 \\ 7\overline{)6,982} \\ -\,63 \\ \hline 68 \\ -\,63 \\ \hline 52 \\ -\,49 \\ \hline 3 \end{array}$$

6. The quotient is 997 R3.

6-6 Long Division with Two-Digit Divisors (pp. 269–274)

Find each quotient. Estimate first. Show the remainder.

41 $1,078 \div 21 =$

_____

42 $901 \div 18 =$

_____

43 $544 \div 47 =$

_____

44 $359 \div 70 =$

_____

Example 7

Find $866 \div 32$. Show the remainder.

1. Estimate. $900 \div 30 = 10$

2. Rewrite the problem in vertical format.

3. Look at the first two digits in the dividend. Think: What number multiplied by 32 is is close to 86? **2** Multiply. Subtract.

4. Think: What number multiplied by 22 is close to 226? **7** Multiply. Subtract.

5. What is the remainder? **2**

6. The quotient is 27 R2.

$$\begin{array}{r} 27 \\ 32\overline{)866} \\ -\,64 \\ \hline 226 \\ -\,224 \\ \hline 2 \end{array}$$

Find each quotient. If the quotient is not possible, write *not possible*.

1 $0 \div 7 =$ _____

2 $33 \div 33 =$ _____

3 $450 \div 10 =$ _____

4 $24 \div 1 =$ _____

5 $14 \div 0 =$ _____

6 $90 \div 10 =$ _____

Use a number line to find the quotient.

7 $16 \div 4 =$ _____

8 $20 \div 5 =$ _____

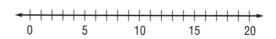

Use a related multiplication fact to find the quotient.

9 $108 \div 12 =$ _____

10 $40 \div 8 =$ _____

_____

_____

11 $77 \div 11 =$ _____

12 $54 \div 9 =$ _____

_____

_____

Use a model to find the quotient.

13 $36 \div 9 =$ _____

14 $16 \div 8 =$ _____

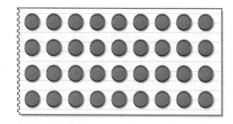

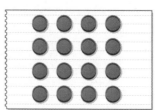

Find each quotient. Show the remainder if there is one.

15 $58 \div 8 =$ _____

16 $2,525 \div 18 =$ _____

17 $120 \div 5 =$ _____

18 $13,473 \div 27 =$ _____

19 $2,275 \div 8 =$ _____

20 $57 \div 9 =$ _____

Solve.

21 COMMUNITY SERVICE   The sixth grade is collecting plastic bottles to donate to a charity that will trade them in for cash. The students have 5 weeks to collect. If their goal is to collect 500 bottles, how many should they collect each week?

_____

22 TRAVEL   A boat travels 384 miles in 24 hours. What is the distance the boat traveled in 1 hour?

_____

23 PHOTOS   Gregory's digital camera holds 43 pictures. Suppose he takes the same number of pictures each day. How many pictures can he take on a 4-day vacation? How many pictures will he have left?

_____

_____

24 READING   Emma read 124 hours in January. If she read an equal number of hours each day, how many hours did she read each day?

_____

Correct the mistakes.

25 TEAMS   During Marcel's family reunion, they want to have a softball tournament. There are 86 family members that are interested in playing. There are 6 teams. Marcel used long division to find out how many members would be on each team. He said that all teams would be even. What did he do wrong?

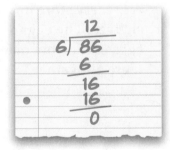

_____

_____

_____

Chapter 6 Test   281

Copyright © Glencoe/McGraw-Hill, a division of The McGraw-Hill Companies, Inc.

Test Practice

Choose the best answer and fill in the corresponding circle on the sheet at right.

1 A _____ is left over after one whole number is divided by another.

    A quotient   **C** remainder

    B dividend   **D** divisor

2 A total of 180 students went on a field trip. There were 3 buses. If each bus had the same number of students on it, how many students were on each bus?

    A 50   **C** 90

    B 60   **D** 210

3 $73 \div 2 =$

    A 34   **C** 35 R1

    B 34 R 7   **D** 36 R1

4 A grocery store has 636 eggs on its shelves. The eggs are in cartons of 12 eggs each. How many cartons are there altogether?

    A 51

    B 52

    C 53

    D 54

5 Which of the following is not possible?

    A $12 \div 1$ **C** $12 \div 0$

    B $0 \div 12$ **D** $12 \div 12$

6 Which equation can be used to check $90 \div 10$?

    A $10 \times 9 = 90$ **C** $90 \div 90 = 1$

    B $10 \times 10 = 100$ **D** $90 \div 1 = 90$

7 Which division sentence represents the array?

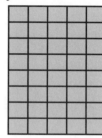

    A $40 \div 4 = 8$

    B $40 \div 4 = 10$

    C $40 \div 8 = 8$

    D $40 \div 8 = 5$

8 $388 \div 24 =$

    A 16

    B 16 R4

    C 16 R5

    D 17

9 The music teacher wants to divide the band students into 8 equal groups. If there are 128 total students in band, which grouping will work?

 A 8 groups of 12 students

 B 9 groups of 15 students

 C 8 groups of 16 students

 D 11 groups of 10 students

10 A square has a perimeter of 36 meters. What is the length of a side of the square?

 A 6 m   C 12 m

 B 9 m   D 15 m

11 Which of the following shows the estimated value of 277 ÷ 3?

 A $210 \div 3 = 70$

 B $300 \div 3 = 100$

 C $300 \div 30 = 10$

 D $350 \div 10 = 35$

12 Adriano is baking cookies. He wants to make 153 large cookies to sell for a school fundraiser. If he can make 9 cookies per batch, how many batches will he need to bake?

 A 14 batches   C 16 batches

 B 15 batches   D 17 batches

ANSWER SHEET

Directions: Fill in the circle of each correct answer.

1 Ⓐ Ⓑ Ⓒ Ⓓ
2 Ⓐ Ⓑ Ⓒ Ⓓ
3 Ⓐ Ⓑ Ⓒ Ⓓ
4 Ⓐ Ⓑ Ⓒ Ⓓ
5 Ⓐ Ⓑ Ⓒ Ⓓ
6 Ⓐ Ⓑ Ⓒ Ⓓ
7 Ⓐ Ⓑ Ⓒ Ⓓ
8 Ⓐ Ⓑ Ⓒ Ⓓ
9 Ⓐ Ⓑ Ⓒ Ⓓ
10 Ⓐ Ⓑ Ⓒ Ⓓ
11 Ⓐ Ⓑ Ⓒ Ⓓ
12 Ⓐ Ⓑ Ⓒ Ⓓ

Success Strategy

Read the entire question before looking at the answer choices. Watch for words like *not* that change the whole question.

STOP

Ratios, Rates, and Unit Rates

How do you determine speed?

Speed is measured by comparing two numbers. It is the ratio of the distance traveled to the time in which it is done. Speed can also be thought of as a rate, for example, miles per hour.

Math Online > Are you ready for Chapter 7? Take the Online Readiness Quiz at *glencoe.com* to find out.

STEP **2** Preview

Get ready for Chapter 7. Review these skills and compare them with what you will learn in this chapter.

What You Know	What You Will Learn
You know how to write fractions to represent parts of a group.  $\frac{3}{5}$ of the birds are bluebirds.	*Lesson 7-1* **Ratios** are a way to compare numbers. A common way to write a ratio is as a fraction in simplest form. There are 3 bluebirds for every 5 birds. The ratio of bluebirds to birds is $\frac{3}{5}$. The ratio of bluebirds to red birds is $\frac{3}{2}$.
You know how to simplify fractions. **Example:** $\frac{150 \div 30}{30 \div 30} = \frac{5}{1} = 5$ **TRY IT** Simplify each fraction. 1 $\frac{36}{12} = $ _____ 2 $\frac{117}{13} = $ _____ 3 $\frac{90}{6} = $ _____	*Lesson 7-3* A **rate** is a ratio that compares different units. When a rate has a denominator of 1, it is a **unit rate**. A teacher hands out 150 pencils to a class of 30 students. Each student gets the same number of pencils. How many pencils for each student? $\frac{150 \text{ pencils}}{30 \text{ students}} = \frac{5 \text{ pencils}}{1 \text{ student}} \rightarrow$ Each student gets 5 pencils.
You know that fractions represent parts of a whole or parts of a set.  The fraction that represents the number of pennies in the set is $\frac{2}{7}$.	*Lesson 7-4* Suppose you have the coins shown at left in your pocket. You pull a coin out of your pocket without looking. The chances are that it would be a penny are $\frac{2}{7}$. The chances of something happening is the **probability** that it will happen.

KEY Concept

A **ratio** is a comparison of two quantities by division. Ratios can compare a part to a part, a part to a whole, or a whole to a part.

Ratios can be written as **fractions**.

Compare part to part

There are 4 red pens to 8 blue pens. The ratio is $\frac{4}{8}$. Other ways to write the ratio:

4 to 8   4:8

Compare part to whole

There are 4 red pens for every 12 pens. The ratio is $\frac{4}{12}$. Other ways to write the ratio:

4 to 12   4:12

Example 1

Write the ratio that compares the number of circles to the number of triangles. Explain the meaning of the ratio.

1. Write the ratio with the number of circles in the numerator and the number of triangles in the denominator.

$$\frac{4}{6} \begin{array}{l} \leftarrow \text{circles} \\ \leftarrow \text{triangles} \end{array}$$

2. The numerator and denominator have a common factor of 2.

Write the fraction in simplest form.

$$\frac{4}{6} = \frac{4 \div 2}{6 \div 2} = \frac{2}{3}$$

3. The ratio of the number of circles to the number of triangles is written as $\frac{2}{3}$, 2 to 3, or 2:3.

4. The ratio means *for every 2 circles, there are 3 triangles.*

YOUR TURN!

Write the ratio that compares the number of circles to the total number of figures. Explain the meaning of the ratio.

1. Write the ratio.

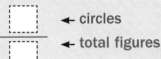 ← circles

── ← total figures

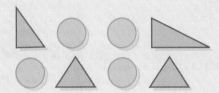

2. The numerator and denominator have a common factor of _____.
   Write the fraction in simplest form.

$$\frac{\square}{\square} = \frac{\square \div \square}{\square \div \square} = \frac{\square}{\square}$$

3. Write the ratio of the number of circles to the number of figures.

_____

4. What does the ratio mean? _____

_____

Example 2

Write the ratio as a fraction in simplest form.

4 red apples out of 10 total apples

1. Write the ratio with the number of red apples in the numerator and the total number of apples in the denominator.

   $\frac{4}{10}$

2. The numerator and denominator have a common factor of 2. Divide each by 2 to write the fraction in simplest form.

   $\frac{4}{10} = \frac{4 \div 2}{10 \div 2} = \frac{2}{5}$

YOUR TURN!

Write the ratio as a fraction in simplest form.

15 DVDs to 18 CDs

1. Write the ratio.

   $\frac{\square}{\square}$

2. The numerator and denominator have a common factor of _____. Write the fraction in simplest form.

   $$\frac{\square}{\square} = \frac{\square \div \square}{\square \div \square} = \frac{\square}{\square}$$

GO ON

Example 3

Write the ratio of the **width** to the **length** in the rectangle as a fraction in simplest form.

3 cm
12 cm

1. Write the ratio as a fraction with the **width** over the **length**.

$$\frac{3}{12}$$

2. The numerator and denominator have a common factor of 3. Divide each by 3 to write the fraction in simplest form.

$$\frac{3}{12} = \frac{3 \div 3}{12 \div 3} = \frac{1}{4}$$

YOUR TURN!

Write the ratio of the **length** to the **width** in the rectangle as a fraction in simplest form.

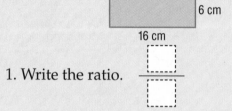
6 cm
16 cm

1. Write the ratio.

2. The numerator and denominator have a common factor of _____. Write the fraction in simplest form.

Who is Correct?

Write the ratio as a fraction in simplest form.   8 plates to 4 cups

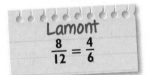
Lamont
$$\frac{8}{12} = \frac{4}{6}$$

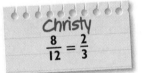
Christy
$$\frac{8}{12} = \frac{2}{3}$$

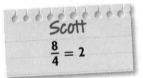
Scott
$$\frac{8}{4} = 2$$

Circle correct answer(s). Cross out incorrect answer(s).

▶ Guided Practice

Use the diagram to write each ratio as a fraction in simplest form.

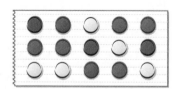

1 The number of red counters to the number of blue counters is _____.

2 The number of red counters to the total number of counters is _____.

3 The number of blue counters to the total number of counters is _____.

Step by Step Practice

4 **FISH**   An aquarium has 7 guppies, 3 angelfish, 5 butterflyfish, and 6 boxfish. Write the ratio of each type of fish to the total number of fish in the aquarium. Write each as a fraction in simplest form.

Step 1   The total number of fish is _____.

This will be the _____ in the fraction.

Step 2   Write a ratio for the number of guppies to the total

number of fish. _____

Write the fraction in simplest form.

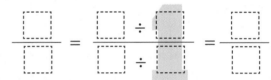

Step 3   Write a ratio for the number of angelfish to the total

number of fish. _____

Write the fraction in simplest form.

Step 4   Write a ratio for the number of butterflyfish to the total

number of fish. _____

Step 5   Write a ratio for the number of boxfish to the total

number of fish. _____

Write the fraction in simplest form.

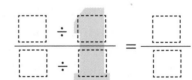

Write each ratio as a fraction in simplest form.

5   In a box of granola bars, there are 6 cinnamon bars and 3 almond bars. Write the ratio of almond bars to cinnamon bars.

almond granola bars   →

cinnamon granola bars   →

GO ON

6 In a sports equipment closet, there are 10 softballs, 4 basketballs, and 3 soccer balls. Write the ratio of soccer balls to the total number of balls. _____

7 In a classroom, there are 24 students and 5 computers. Write the ratio of students to computers. _____

8 In a bag of 18 marbles, there are 16 that are *not* white. Write the ratio of white marbles to nonwhite marbles. _____

Step by Step Problem-Solving Practice

Solve.

Problem-Solving Strategies
☐ Look for a pattern.
☐ Guess and check.
☐ Act it out.
☑ Solve a simpler problem.
☐ Work backward.

9 **AGES** Charles is 12 years old, and his sister Tracy is 8 years old. In four years, what will be the ratio of Charles' age to Tracy's age?

Understand Read the problem. Write what you know. Charles is _____ years old. Tracy is _____ years old. In 4 years, Charles will be _____ years old, and Tracy will be _____ years old.

Plan Pick a strategy. One strategy is to solve a simpler problem.

Solve First, write the ratio of their ages in four years. _____

To write the ratio in simplest form, divide the numerator and denominator by a common factor. Divide the numerator and denominator by 2.

$$\frac{\Box}{\Box} = \frac{\Box \div \Box}{\Box \div \Box} = \frac{\Box}{\Box}$$

Is there still a common factor? _____ Divide the numerator and denominator by 3.

$$\frac{\Box}{\Box} = \frac{\Box \div \Box}{\Box \div \Box} = \frac{\Box}{\Box}$$

Is there still a common factor? _____

Write the ratio in simplest form. _____

Check Use counters or a diagram to model the problem.

10 **SOFTBALL**   Isabel played all last summer on the softball team.
Their team record for the summer was 20 wins and 6 losses. Write
the ratio of wins to losses in a fraction in simplest form.
Check off each step.

_____ Understand: I underlined key words.

_____ Plan: To solve the problem, I will _____ .

_____ Solve: The answer is _____ .

_____ Check: I checked my answer by _____ .

11 **TENNIS**   Jimmy and Tenisha played 20 sets of tennis. Jimmy won
12 of them. Write a ratio of Jimmy's wins to the total number of sets
in simplest form.

_____

12 **Reflect**   What is a ratio? Explain using examples.

_____

_____

▶ Skills, Concepts, and Problem Solving

Use the diagram to write each ratio as a fraction in simplest form.

13 nickels and dimes to pennies and quarters _____

14 coins that that are *not* quarters to total number

    of coins _____

15 nickels to dimes and pennies _____

Write each ratio as a fraction in simplest form.

16 Stephanie jogged 10 miles in 100 minutes. _____

17 There are 12 puppies to 15 kittens at the pet store. _____

18 Russell had 6 hits out of 10 at bats. _____

Write the ratio of length to width in each rectangle as a fraction in simplest form.

19 [rectangle: 12 m, 50 m] _____

20 [rectangle: 11 mm, 7 mm] _____

Refer to the table below to answer Exercises 21–23.

CELL PHONES A new wireless company is offering special packages for the first month of business for different models of cell phones.

21 Which cell phone model was sold the most?

_____

What was the ratio to the total number of cell phone models sold?

_____

22 Compare the number of global models sold to the number of business models sold.

_____

23 Write the ratio for the number of education model cell phones sold to the total number of cell phones sold in three ways.

_____

Model	Number of Cell Phones Sold
Business	18
Education	25
Entertainment	45
Global	12

Vocabulary Check **Write the vocabulary word that completes each sentence.**

24 A(n) _____ compares two quantities.

25 A(n) _____ can represent a part of a whole, part of a set, or a ratio.

26 **Writing in Math** Write the ratio of *7 rulers out of a total of 10 rulers* four different ways.

_____

STOP

Equivalent Ratios

KEY Concept

A **ratio table** is a table with columns filled with pairs of numbers that have the same ratio.

Suppose there are exactly 3 boys for every 5 students. The **ratio** is $\frac{3}{5}$. Multiplying the original ratio by an **equivalent form of one** produces an **equivalent ratio**. This can be done many times, producing a table of fractions that share the same original ratio.

Number of Boys	3	6	9	12
Number of Students	5	10	15	20

The table is completed by multiplying by equivalent forms of one.

$$\frac{3 \times 2}{5 \times 2} = \frac{6}{10} \qquad \frac{3 \times 3}{5 \times 3} = \frac{9}{15} \qquad \frac{3 \times 4}{5 \times 4} = \frac{12}{20}$$

VOCABULARY

equivalent forms of one
different expressions that represent the same number

equivalent ratios
ratios that have the same value

fraction
a number that represents part of a whole or part of a set

ratio
a comparison of two quantities by division

ratio table
a table with columns filled with pairs of numbers that have the same ratio

Example 1

Fill in the blanks and complete the ratio table.

1. The original ratio is $\frac{2}{7}$.

2. Complete the table. Multiply by equivalent forms of one.

		×2	×3	×4	×5	×6
Numerator	2	4	6	8	10	12
Denominator	7	14	21	28	35	42

3. The missing ratio is $\frac{12}{42}$.

GO ON

Fill in the blanks and complete the ratio table.

1. The original ratio is _____.

2. Complete the table. Multiply by equivalent forms of one.

		× 3	× 5	× ___	× ___	× ___
Numerator	5	15	25	35	45	
Denominator	6	18	30	42	54	

| | | × 3 | × 5 | × ___ | × ___ | × ___ |

3. The missing ratio is _____.

Example 2

Fill in the blanks and complete the ratio table.

1. The original ratio is $\frac{48}{80}$.

2. Complete the table. Divide by equivalent forms of one.

		÷ 2	÷ 4	÷ 8	÷ 16
Numerator	48	24	12	6	3
Denominator	80	40	20	10	5

| | | ÷ 2 | ÷ 4 | ÷ 8 | ÷ 16 |

3. The missing ratio is $\frac{3}{5}$.

Fill in the blanks and complete the ratio table.

1. The original ratio is _____.

2. Complete the table. Divide by equivalent forms of one.

		÷ 10	÷ 20	÷ ___	÷ ___
Numerator	120	12	6	4	
Denominator	240	24	12	8	

| | | ÷ 10 | ÷ 20 | ÷ ___ | ÷ ___ |

3. The missing ratio is _____.

Who is Correct?

List two equivalent ratios of $\frac{4}{7}$.

Kelvin

$\frac{4}{7} = \frac{2}{3} = \frac{16}{28}$

Amir

$\frac{4}{7} = \frac{8}{14} = \frac{12}{21}$

Vanessa

$\frac{4}{7} = \frac{7}{4} = \frac{14}{8}$

Circle correct answer(s). Cross out incorrect answer(s).

 ## Guided Practice

Fill in the blanks and complete the ratio tables.

1

	× ___	× ___	× ___	
Numerator	1	4	8	
Denominator	4	16	32	

× ___   × ___   × ___

2

	÷ ___	÷ ___	÷ ___	
Numerator	210	42	21	
Denominator	360	72	36	

÷ ___   ÷ ___   ÷ ___

Step by Step Practice

3 Fill in the blanks and complete the ratio table.

Numerator	1	2	3	
Denominator	3	6	9	

Step 1   The original ratio is _____.

Step 2   Complete the table. Multiply by equivalent forms of one.

Step 3   The missing ratio is _____.

GO ON

Fill in the blanks and complete the ratio tables.

4

Numerator	1	2	3	
Denominator	6	12	18	

5

Numerator	2	4	6	
Denominator	7	14	21	

Step by Step Problem-Solving Practice

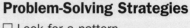

Problem-Solving Strategies
- ☐ Look for a pattern.
- ☐ Guess and check.
- ☐ Act it out.
- ☑ Use a table.
- ☐ Work backward.

6 **HEALTH** Marita is writing a report on the amount of sugar in cola. She knows 1 can of cola has 39 grams of sugar. She wants to find the amount of sugar in different numbers of cans of cola.

Understand Read the problem. Write what you know.

The base ratio of sugar to cola is _____.

Plan Pick a strategy. One strategy is to use a table.

Solve Find a pattern to find the equivalent forms of one that determine the equivalent ratios.

	× 2	× 3	× ___	× ___
Grams (g) of sugar	39		156	195	
Cans of cola	1	2	3	4	5
	× 2	× 3	× ___	× ___

Complete the table.

Check Work backward using division to check.

7 **MUSIC** The Grooves music store is having a sale on CDs. You can buy 3 CDs for $10. Complete the ratio table to show how much it would cost to buy 9 and 12 CDs.

Number of CDs	3	6	9	12
Cost ($)	10	20		

_____ Understand: I underlined key words.

_____ Plan: To solve the problem, I will _____.

_____ Solve: The answer is _____.

_____ Check: I checked my answer by _____.

8 [Reflect] Name 3 equivalent forms of one for $\frac{2}{3}$.

▶ Skills, Concepts, and Problem Solving

Fill in the blanks and complete the ratio tables.

9

Numerator	3	15	21	
Denominator	2	10	14	

10

Numerator	200	100	50	
Denominator	48	24	12	

11

Numerator	7		21	28
Denominator	15	30		

12

Numerator	150	50		15
Denominator	300	100	60	

13 **TEXTILES** Stefan's company makes waterproof tents. How many yards of fabric must Stefan have on hand to fill the following tent orders? How much fabric must he have for one tent?

Yards of fabric	280	560	1120	2240	4480
Tents	32	64			

14 **GAS MILEAGE** Kaya is planning a road trip. Complete the table to show the ratio of miles to be traveled to gallons of gasoline.

Miles to be traveled	48	144	240	336	432
Gallons of gasoline	2				

GO ON

Vocabulary Check   **Write the vocabulary word that completes each sentence.**

15 Different expressions that represent the same number are called

_____.

16 A(n) _____ is a table with columns filled with pairs of numbers that have the same ratio.

17 Ratios that have the same value are _____.

18 **Writing in Math**   How do you use equivalent forms of one to complete a ratio table?

_____

_____

▶ Spiral Review

Write the ratio of width to length in each rectangle as a fraction in simplest form. (Lesson 7-1, p. 286)

19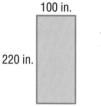
100 in.
220 in.
_____

20
4 m
16 m
_____

21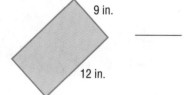
9 in.
12 in.
_____

22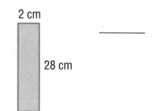
2 cm
28 cm
_____

STOP

Progress Check 1 (Lessons 7-1 and 7-2)

Use the diagrams to write each ratio as a fraction in simplest form.

1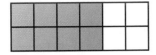

shaded to unshaded squares _____

2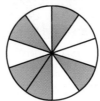

unshaded parts to total parts _____

Write the ratio of width to length in each rectangle as a fraction in simplest form.

3 4 km _____

5 km

4 10 in. _____

2 in.

Write each ratio as a fraction in simplest form.

5 19 out of 133 girls had green eyes _____

6 5 long-haired cats out of 12 cats _____

Complete each ratio table.

7

Numerator	9	27	36	45
Denominator	13			65

8

Numerator	360	180		60	
Denominator	48		12		6

Solve.

9 **SHOPPING** Glenda went to the grocery store for grapes. The grapes were $6 for 3 pounds. Write the ratio of cost to pounds in simplest form.

_____

10 **COST** The student council treasurer made a ratio table showing the cost of purchasing 7, 13, 24, and 35 cases of fruit bars. Use the ratio table to find the cost of buying one case of fruit bars.

Cases		7	13	24	35
Cost in dollars		91	169	312	455

Rates

KEY Concept

Copyright © Glencoe/McGraw-Hill, a division of The McGraw-Hill Companies, Inc.

A **rate** is a **ratio** of two measurements having different units.

300 miles in 5 days → $\dfrac{300 \text{ miles}}{5 \text{ days}}$

> The units *miles* and *days* are different.

When a rate is simplified so that it has a denominator of 1 unit, it is called a **unit rate**.

50 miles per hour → $\dfrac{50 \text{ miles}}{1 \text{ hour}}$

> The denominator is 1 unit.

Unit cost is the cost of a single item or unit of measurement.

The cost of a 12-ounce jar of jam is $2.49.

$\dfrac{2.49}{12}$ → $12\overline{)2.49}$ → about 0.21 → $\dfrac{21 \text{ cents}}{1 \text{ ounce}}$

> The unit cost is 21 cents per ounce.

VOCABULARY

rate
a ratio comparing two quantities with different kinds of units

ratio
a comparison of two numbers by division

unit cost
the cost of a single item

unit rate
a rate that describes how many units of the first type of quantity are equal to 1 unit of the other type of quantity

Rates are often written using abbreviations, such as 300 mi/5 days, 60 mi/h, or $0.21/oz.

Example 1

Write the rate 50 claps in 5 seconds as a fraction. Find the unit rate.

1. Write the rate as a fraction. $\dfrac{50 \text{ claps}}{5 \text{ seconds}}$

2. Find an equivalent rate with a denominator of 1. The numerator and denominator have a common factor of 5. Divide each by 5.

   $\dfrac{50 \text{ claps} \div 5}{5 \text{ seconds} \div 5} = \dfrac{10 \text{ claps}}{1 \text{ second}}$

3. Name the unit rate.

   10 claps per second or 10 claps/s

YOUR TURN!

Write the rate 90 miles in 2 hours as a fraction. Find the unit rate.

1. Write the rate as a fraction.

   $\dfrac{\boxed{} \text{ miles}}{\boxed{} \text{ hours}}$

2. The numerator and denominator have a common factor of _____.

   $\dfrac{\boxed{} \text{ miles} \div \boxed{}}{\boxed{} \text{ miles} \div \boxed{}} = \dfrac{\boxed{} \text{ miles}}{\boxed{} \text{ hour}}$

3. Name the unit rate.

   _____

Example 2

Find the unit rate for selling 300 tickets in 6 days. Use the unit rate to find the number of tickets sold in 5 days.

1. Write the rate as a fraction.

   $$\frac{300 \text{ tickets}}{6 \text{ days}}$$

2. Find an equivalent rate with a denominator of 1.

   Divide the numerator and denominator by 6.

   $$\frac{300 \div 6}{6 \div 6} = \frac{50}{1}$$

3. The unit rate is 50 tickets/day.

4. To find how many tickets will be sold at this rate in 5 days, multiply the numerator and denominator by 5.

   $$\frac{50 \text{ tickets} \times 5}{1 \text{ day} \times 5} = \frac{250 \text{ tickets}}{5 \text{ days}}$$

At this rate, 250 tickets will be sold in 5 days.

YOUR TURN!

Find the unit rate for traveling 165 feet in 15 seconds. Use the unit rate to find the number of feet traveled in 120 seconds.

1. Write the rate as a fraction.

   $$\frac{\boxed{} \text{ ft}}{\boxed{} \text{ s}}$$

2. Divide the numerator and denominator by _____.

   $$\frac{\boxed{} \text{ ft} \div \boxed{}}{\boxed{} \text{ s} \div \boxed{}} = \frac{\boxed{} \text{ ft}}{\boxed{} \text{ s}}$$

3. The unit rate is _____/_____.

4. Multiply the numerator and denominator by _____.

   $$\frac{\boxed{} \text{ ft} \times \boxed{}}{\boxed{} \text{ s} \times \boxed{}} = \frac{\boxed{} \text{ ft}}{\boxed{} \text{ s}}$$

At this rate, _____ feet will be traveled in 120 seconds.

Example 3

Ms. Jackson bought a box of greeting cards for $5.75. The box contains 12 cards. Find the unit cost to the nearest cent.

1. Write the rate as a fraction.
   $$\frac{\$5.75}{12 \text{ cards}}$$

2. Divide the numerator by the denominator.

3. $0.47\frac{11}{12}$ rounded to the nearest cent is $0.48.

$$\begin{array}{r} 0.47 \\ 12\overline{)5.75} \\ -48 \\ \hline 95 \\ -84 \\ \hline 11 \end{array}$$

Each card costs about $0.48.

YOUR TURN!

Mr. Tucker bought a box of oranges for $12.50. The box contains 15 oranges. Find the unit cost to the nearest cent.

1. Write the rate as a fraction.

   _____

2. Divide the numerator by the denominator.

3. The unit cost rounded to the nearest cent is

   _____.

Each orange costs about _____.

GO ON

Who is Correct?

Gavin can drive 220 miles on 8 gallons of gas. Find the unit rate. Use the unit rate to find the number of miles Gavin can drive on 64 gallons of gas.

Miguel

$$\frac{220 \div 8}{8 \div 8} = \frac{27}{1}$$

Unit rate = 27 mi/gal;
27 × 64 = 1,728 mi

Imani

$$\frac{220 \div 8}{8 \div 8} = \frac{27.5}{1}$$

Unit rate = 27.5 mi/gal;
27.5 × 64 = 1,760 mi

Xavier

$$\begin{array}{r} 27.5 \\ 8\overline{)220.0} \\ \underline{16} \\ 60 \\ \underline{56} \\ 40 \\ 40 \end{array}$$

Unit rate is 27.5 mi/gal
27.5 mi/gal × 64 =
1,760 mi on 64 gal

Circle correct answer(s). Cross out incorrect answer(s).

▶ Guided Practice

Write each rate as a fraction. Find each unit rate.

1 12 books in 5 days

_____

2 168 miles in 3 hours

_____

3  $2.00

_____

4 140 words in 4 minutes

_____

Find each unit rate. Use the unit rate to find the unknown amount.

5 $5 for 4 books; □ dollars for 15 books _____

6 150 feet in 8 seconds; □ feet in 14 seconds _____

7 12 hours for 5 classes; □ hours for 4 classes _____

8 5 pounds for 8 people; □ pounds for 20 people _____

Step (by) Step Practice

9 Use the table to find which box of macaroni has the lowest unit cost. Round to the nearest cent.

Step 1 Find the unit cost of a 12-oz package.

$$\frac{0.90}{12} \rightarrow 12\overline{)0.90} \rightarrow \text{about } \$\_\_\_\_\_\text{/oz}$$

Box Size	Price
12 oz	$0.90
16 oz	$1.12
32 oz	$1.95

Step 2 Find the unit cost of a 16-oz package.

$$\frac{}{} \rightarrow)\overline{} \rightarrow \$\_\_\_\_\_\text{/oz}$$

Round to the nearest cent means the nearest hundredth.

Step 3 Find the unit cost of a 32-oz package.

$$\frac{}{} \rightarrow)\overline{} \rightarrow \text{about } \$\_\_\_\_\_\text{/oz}$$

Step 4 Which package costs the least per ounce?

_____

Which product has the lowest unit cost? Round to the nearest cent.

10 a 12-oz juice bottle for $0.75 or a 24-oz juice bottle for $1.95

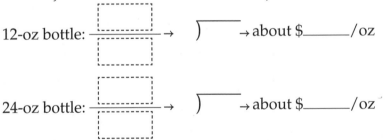

12-oz bottle: $\dfrac{}{} \rightarrow)\overline{} \rightarrow \text{about } \$\_\_\_\_\_\text{/oz}$

24-oz bottle: $\dfrac{}{} \rightarrow)\overline{} \rightarrow \text{about } \$\_\_\_\_\_\text{/oz}$

The _____ juice bottle costs less per ounce.

11 50-count vitamins for $5.49, 100-count vitamins for $8.29, or 150-count vitamins for $12.75 _____

12 a 16-oz bag of apples for $2.99, a 32-oz bag of apples for $3.99, or a 48-oz bag of apples for $5.49 _____

13 a 6-pack of yogurt for $1.99, or a 12-pack of yogurt for $3.50

_____

14 4 shirts for $24.85, or 7 shirts for $49.49 _____

Step by Step Problem-Solving Practice

Solve.

15 BIRDS The American robin can travel 32 miles in a 20-hour flight. The grey-cheeked thrush can travel 33 miles in 6 hours. Which bird flies at a faster rate?

Understand Read the problem. Write what you know.

The American Robin can travel
_____ miles in a _____-hour flight.

The Grey-Cheeked Thrush can travel
_____ miles in a _____-hour flight.

Plan Pick a strategy. One strategy is to solve a simpler problem. Find each unit rate.

Solve Write each rate as a fraction. Find an equivalent rate with a denominator of 1.

Unit Rate of the American Robin

$$\frac{32 \text{ miles} \div }{20 \text{ hours} \div } = \frac{\text{miles}}{1 \text{ hour}}$$

American Robin

Unit Rate of the Grey-Cheeked Thrush

$$\frac{33 \text{ miles} \div }{6 \text{ hours} \div } = \frac{\text{miles}}{1 \text{ hour}}$$

Compare the unit rates for each bird.

_____ miles/hour < _____ miles/per hour

The _____ flies at a faster rate.

Grey-Cheeked Thrush

Check Does the answer make sense? Work backward to check.

16 WAGES While working at a gardening center for the summer, Herman earned $780 in 12 weeks. Find a unit rate to describe his weekly wages. Check off each step.

_____ Understand: I underlined key words.

_____ Plan: To solve the problem, I will _____.

_____ Solve: The answer is _____.

_____ Check: I checked my answer by _____.

17 POPULATION The population of Flordia is about 18.1 million people. Its land area is approximately 54,252 square miles. Find the population per square mile.

_____

18 Reflect Explain the difference between a rate and ratio. What is the difference between unit rate and unit cost?

_____

_____

▶ Skills, Concepts, and Problem Solving

Write each rate as a fraction. Find each unit rate.

19 9 feet every 10 seconds

20 21 hits out of 40 at bats

_____

21 6 pancakes in 4 minutes

22 9 feet in 12 years

_____

Find each unit rate. Use the unit rate to find the unknown amount.

23 150 feet in 8 seconds; ☐ feet in 14 seconds _____

24 9 yards in 3 plays; ☐ yards for 4 plays _____

25 $30 for 16 ounces; ☐ dollars for 6 ounces _____

26 50 meters every 8 seconds; ☐ meters for 20 seconds _____

Which product has the lower unit cost? Round to the nearest cent.

27 4-pack of tissues for $3.39, or 16-pack of tissues for $14.75 _____

28 40-oz can of soup for $4.49, or 25-oz can of soup for $2 _____

29 32-oz shampoo bottle for $6, or 8-oz shampoo bottle for $1.75 _____

30 12 golf balls for $9, or 10 golf balls for $8.50 _____

GO ON

Solve.

31 **HEART RATE** The heart of a hamster beats about 900 times in
2 minutes, while the heart of a guinea pig beats about
1,200 times in 4 minutes. The heartbeat of a rabbit is about
1,025 beats in 5 minutes. Which animal's heart beats the
most times in one hour? Explain.

_____

32 **POPULATION** Which country has the lower
population per square mile? Explain.

_____

_____

Country	Population	Area in sq miles
Japan	128,000,000	377,900
United Kingdom	59,700,000	243,000

Vocabulary Check **Write the vocabulary word that completes each sentence.**

33 A ratio of two measurements or amounts of different
units, where the second amount is 1 is a(n)

_____.

34 The cost of a single item or unit is the _____.

35 **Writing in Math** Which of the following statements are sometimes,
always, or never true? Give an example or counterexample
to illustrate.

      A ratio is a rate.   A rate is a ratio.

_____

_____

 Spiral Review

**Use the diagram at the right to write each ratio
as a fraction in simplest form.** (Lesson 7-1, p. 286)

36 The number of red counters to the number of blue counters is _____.

37 The number of red counters to the total number of counters is _____.

38 The number of blue counters to the total number of counters is _____.

Probability as a Ratio

KEY Concept

Probability is a number that measures the chance of an event happening. The probability of an event is a **ratio** that compares the number of favorable outcomes to the number of possible **outcomes**. The probability of an **event** is written as P(event).

Suppose you roll a number cube.

P(even number) $= \dfrac{\text{number of favorable outcomes}}{\text{total number of outcomes}}$

The even numbers are 2, 4, and 6. $= \dfrac{\text{number of even numbers}}{\text{total number of outcomes}} = \dfrac{3}{6} = \dfrac{1}{2}$

The ratio of the even numbers to the total numbers is $\dfrac{3}{6}$ or $\dfrac{1}{2}$. Notice the ratio and the probability are the same.

Probability can also be written as a decimal or as a percent.

$\dfrac{1}{2}$ 0.50   50%

The probability, $\dfrac{1}{2}$, means that you can expect to roll an even number 1 out of every two times or 50% of the time.

The probability of an event can be 0, 1, or any number between 0 and 1.

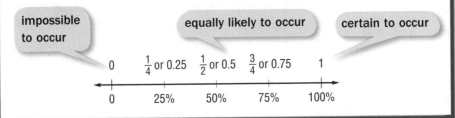

VOCABULARY

event
  a set of outcomes

outcomes
  the possible results of a probability event

probability
  a number between 0 and 1 that measures the likelihood of an event happening

ratio
  a comparison of two numbers by division; the ratio of 2 to 3 can be stated as 2 out of 3, 2 to 3, 2:3, or $\dfrac{2}{3}$

When probability equals 0, the event is **impossible**. For example, the probability of rolling a 7 on a number cube is 0.

When probability equals 1, the event is **certain**. For example, the probability of rolling a natural number that is 6 or less is 1.

The probability that one event does **not** occur is equal to $1 - P$(event does occur).

GO ON

Example 1

Use the spinner to find the probability of spinning 6. Write the probability as a fraction in simplest form. Explain the probability.

1. Count the number of sections labeled 6. Write this number in the numerator.

2. Count the total number of sections. Write this number in the denominator.

$$P(6) = \frac{1}{8}$$

3. The ratio is $\frac{1}{8}$. It is already in simplest form.

4. The probability of $\frac{1}{8}$ means that 1 out of every 8 spins should be a 6.

YOUR TURN!

Use the spinner in Example 1 to find the probability of spinning an odd number. Write the probability as a fraction in simplest form. Explain the probability.

1. What number of sections are labeled with an odd number? _____

2. What is the total number of sections? _____

3. $P(\text{odd number}) = \dfrac{\boxed{}}{\boxed{}}$

4. The ratio can be simplified.

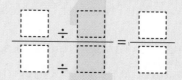

5. The probability of _____ means that _____ out of every _____ spins should be an odd number.

Example 2

A drawer of socks contains 3 pairs of white socks, 3 pairs of blue socks, and 3 pairs of black socks. What is the probability of choosing a pair of black or blue socks if you take 1 pair from the drawer without looking?

1. How many pairs of socks are blue or black? 6

2. How many pairs of socks are in the drawer in all? 9

3. Write a ratio for the $P(\text{black or blue})$. $\frac{6}{9}$

4. Write the fraction in simplest form.
$$\frac{6 \div 3}{9 \div 3} = \frac{2}{3}$$

5. When you take a pair of socks without looking, the probability the socks will be blue or black is two-thirds.

YOUR TURN!

A bowl of fruit has 4 peaches, 5 plums, 5 apples, and 3 oranges. What is the probability that a peach or an orange is selected if you choose a fruit without looking?

1. The number of peaches and oranges is _____.

2. The number of pieces of fruit in the basket is _____.

3. $P(\text{peach or orange}) = \dfrac{\boxed{}}{\boxed{}}$

4. The ratio is already simplified.

5. _____ out of every _____ times you take a piece of fruit from the basket without looking it will be

_____.

Who is Correct?

The ratio of green marbles to the total number of marbles in a bag is $\frac{3}{10}$. What is the probability that when you pick a marble without looking it will not be green?

Oscar
$\frac{3}{10}$

Erika
$\frac{3}{13}$

Cameron
$\frac{7}{10}$

Circle correct answer(s). Cross out incorrect answer(s).

 Guided Practice

Use the spinner to find each probability. Write the probability as a fraction in simplest form.

1 P(multiple of 3) _____

2 P(**not** a multiple of 3) _____

Step by Step Practice

Find the probability. Write the probability as a fraction in simplest form.

3 In a bag there are 5 green chips, 3 yellow chips, 7 blue chips, and 1 red chip. Find the probability of reaching into the bag without looking and not getting a green chip.

Step 1 Count the total number of chips. _____
This is the _____ in the fraction.

Step 2 Count the number of chips in the bag that are green. _____
This is the _____ in the fraction.

Step 3 Write a ratio in simplest form for the P(green). $\frac{\Box}{\Box}$

Step 4 Find the P(not green). $1 - \frac{\Box}{\Box} = \frac{\Box}{\Box}$

GO ON

Find each probability. Write the probability as a fraction in simplest form.

4 In a box of mugs, there are 6 white mugs, 4 blue mugs, and 8 beige mugs. What is the probability that without looking you would choose a mug that is not white?

$$P(\text{not white}) = \frac{\boxed{}}{\boxed{}} = \frac{\boxed{} \div \boxed{}}{\boxed{} \div \boxed{}} = \frac{\boxed{}}{\boxed{}}$$

5 On a serving counter, there are 3 sausage pizzas, 8 cheese pizzas, and 6 pepperoni pizzas. Find the probability of randomly selecting a piece of sausage pizza.

$P(\text{sausage}) = \underline{\hspace{2cm}}$

Find each probability using a number cube. Write the probability as a fraction in simplest form.

6 $P(\text{roll a 3})$ _____

7 $P(\text{roll a 1 or a 6})$ _____

8 $P(\text{roll a number less than 6})$ _____

9 $P(\text{roll a number less than 3})$ _____

10 $P(\text{roll an even number})$ _____

11 $P(\text{roll a 7})$ _____

Find the probability of each event. Write the probability as a fraction in simplest form.

12 You pick a day of the week that begins with the letter *T*. _____

13 You pick a weekend day from the days of the week. _____

14 You pick one of the letters *A*, *F*, or *G* from the alphabet. _____

15 You pick a month that begins with the letter *J*. _____

16 You pick the letter A from the letters in AMERICAN. _____

17 You pick a day of the week that ends in the letter *Y*. _____

Step by Step Problem-Solving Practice

Solve.

Problem-Solving Strategies
- ☐ Draw a diagram.
- ☐ Look for a pattern.
- ☑ Use logical reasoning.
- ☐ Act it out.
- ☐ Solve a simpler problem.

18 **TRAITS**   The ratio of brown-haired students to the total number of students in a class is 22 out of 30. What is the probability, if one student is picked by the teacher without looking, that the student will *not* have brown hair?

Understand   Read the problem. Write what you know.

Out of _____ students, _____ have brown hair.

Plan   Pick a strategy. One strategy is to use logical reasoning.

Solve   Write the ratio of brown-haired students to total students. _____

The probability of choosing a student *who does not have* brown hair is the same as the difference of the ratio of the entire class and the ratio of brown-haired students.

$$\frac{30}{30} - \frac{22}{30} = \frac{8}{30} = \frac{\boxed{}}{15}$$

The probability of choosing a student who does not have brown hair is _____.

Check   Check your answer. The sum of the probability that an event occurs and the probability that the event does not occur is 1. Is the sum of your probabilities equal to one? Explain.

_____

19 **BAGELS**   The probability of buying a dozen bagels and receiving an extra bagel is 2 out of 100. Find the probability of *not* receiving an extra bagel.
Check off each step.

_____ **Understand: I underlined key words.**

_____ **Plan: To solve the problem, I will** _____.

_____ **Solve: The answer is** _____.

_____ **Check: I checked my answer by** _____.

GO ON

20 **MARBLES** The probability of choosing a black marble out of a bag of marbles without looking is $\frac{3}{14}$. What is the probability of *not* picking a black marble? _____

21 **Reflect** How are probability and ratios the same?

_____

_____

▶ Skills, Concepts, and Problem Solving

Use the spinner to find each probability. Write the probability as a fraction in simplest form.

22 P(white) _____

23 P(not white) _____

24 Add your answers to Exercises 22 and 23. What is their sum? _____

Use the basket of fruit to find each probability. Write the probability as a fraction in simplest form.

25 Write the ratio for the number of plums and pears to the total number of fruit. _____

26 Write the ratio for the number of apples to the number of bananas. _____

27 What is the probability of choosing fruit from the basket without looking and getting an apple or a banana? _____

28 What is the probability of choosing fruit from the basket without looking and getting a fruit that is *not* an apple or banana? _____

Find each probability. Write the probability as a fraction in simplest form.

29 7 red hats, 9 green hats, and 4 blue hats; P(blue hat) _____

30 2 small popcorn bags, 5 medium popcorn bags, and 3 large popcorn bags; P(small or large popcorn bags) _____

31 9 fourth graders, 6 second graders, and 2 third graders; P(not a second grader) _____

32 **BUTTONS** The ratio of blue buttons to the total number of buttons in a tin is $\frac{4}{9}$. What is the probability if a button is chosen without looking that the button will *not* be blue? _____

33 **LETTERS** Suppose the letters of the word *mathematics* are placed in a bag. A letter is pulled out without looking. What is the probability that the letter is an *m*? _____

Vocabulary Check **Write the vocabulary word that completes each sentence.**

34 _____ is a number between 0 and 1 that measures the likelihood of an event.

35 A(n) _____ compares two quantities.

36 **Writing in Math** Write an example of a situation in which the probability of an event occurring is 0.

_____

_____

 Spiral Review

Find each unit rate. Use the unit rate to find the unknown rate.
(Lesson 7-3, p. 300)

37 10 feet every 50 seconds; □ feet for 30 seconds _____

38 9 hits out of 36 at bats; □ hits for 44 at bats _____

Solve. (Lesson 7-3, p. 300)

39 **HEALTH** After running in a race, Ciera's heart rate is 111 beats per minute. After running the same race, Adriana's heart rate is 235 beats every 2 minutes. Who has a faster heart rate?

_____

40 **INSECTS** Caterpillar A travels 6 meters in 5 hours. Caterpillar B travels 30 meters in 20 hours. Which caterpillar travels more slowly?

_____

Write each rate as a fraction. Find each unit rate.

1 45 miles in 9 minutes _____

2 3 tons in 75 years _____

Which product has the lowest unit cost? Round to the nearest cent.

3 12-oz can for $1.99, a 16-oz can for $2.50, or a 32-oz can for $3.79

_____

4 9 kiwis for $1.35, 14 kiwis for $2.25, or 20 kiwis for $3.80

_____

Use the spinner to find each probability. Write the probability as a fraction in simplest form.

5 P(blue) _____

6 P(not blue) _____

7 P(red, yellow, or purple) _____

8 P(not green or white) _____

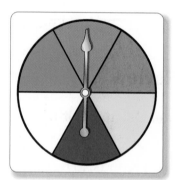

Solve.

9 **SPORTS**   Alaria swam 200 feet in 44 seconds. What is her unit rate?

_____

10 **MOVIES**   Nate surveyed the sixth grade class about their favorite type of movies. What is the probability of the students *not* choosing thrillers as their favorite type of movie? Write the probability as a fraction in simplest form.

_____

Type of Movie	Number of Students
Adventure	40
Comedy	28
Romance	7
Thriller	15

Vocabulary and Concept Check

equivalent forms of one, *p. 293*

equivalent ratios, *p. 293*

event, *p. 307*

fraction, *p. 286*

outcomes, *p. 307*

probability, *p. 307*

rate, *p. 300*

ratio, *p. 286*

ratio table, *p. 293*

unit cost, *p. 300*

unit rate, *p. 300*

Write the vocabulary word that completes each sentence.

1 _____ is a number between 0 and 1 that measures the likelihood of an event happening.

2 A(n)_____ is a ratio of two measurements or amounts made with different units, such as 2 miles in 5 minutes.

3 Different expressions that represent the same number are called _____.

4 A(n) _____ is a comparison of two numbers by division.

5 The probability of a(n) _____ can be written as a fraction.

Write the correct vocabulary term in each blank.

6 $2.73 per gallon

_____

7 65 miles per hour

_____

Lesson Review

7-1 Ratios (pp. 286–292)

Write each ratio as a fraction in simplest form.

8 16 apples to 24 oranges _____

9 5 strawberries to 15 cherries _____

10 12 televisions to 4 radios _____

11 21 pens to 33 pencils _____

Example 1

Write the ratio as a fraction in simplest form.
3 white shirts out of 15 total shirts

1. Write the ratio with the number of white shirts in the numerator and the total number of shirts in the denominator.

$$\frac{3}{15}$$

2. Write the fraction in simplest form.

$$\frac{3 \div 3}{15 \div 3} = \frac{1}{5}$$

7-2 Equivalent Ratios (pp. 293–298)

Fill in the blanks and complete each ratio table.

12

	×2	×___	×___
Numerator	4	8	12	
Denominator	9	18	27	

×2 ×___ ×___

13

	÷5	÷___	÷___
Numerator	30	6	3	
Denominator	90	18	9	

÷5 ÷___ ÷___

Example 2

Fill in the blanks and complete the ratio table.

1. The original ratio is $\frac{1}{6}$.

2. Complete the table. Multiply by equivalent forms of one.

	×3	×6	×9	×12
Numerator	1	3	6	9	12
Denominator	6	18	36	54	72

×3 ×6 ×9 ×12

3. The missing ratio is $\frac{12}{42}$.

Complete each ratio table.

14

Numerator	4	40	36	32
Denominator	21		189	

15

Numerator	11	33		
Denominator	15	45	75	

16

Numerator	320		80	40
Denominator	480			60

17

Numerator	72	24		8
Denominator	144		24	

7-3 Rates and Unit Costs (pp. 300–306)

Write each rate as a fraction. Find each unit rate.

18 100 miles in 4 hours

_____

19 60 gallons in 5 minutes

_____

20 55 meters in 11 seconds

_____

Example 3

Write the rate 80 beats per 10 seconds as a fraction. Find the unit rate.

1. Write the rate as a fraction.

$$\frac{80 \text{ beats}}{10 \text{ seconds}}$$

2. Find an equivalent rate with a denominator of 1.

$$\frac{80 \text{ beats} \div 10}{10 \text{ seconds} \div 10} = \frac{8 \text{ beats}}{1 \text{ second}}$$

3. Name the unit rate.

8 beats per second or 8 beats/s

7-4 Probability as a Ratio (pp. 307–313)

21 Use the spinner in Example 4 to find the probability of spinning a 3 or a 4. Write the probability as a fraction in simplest form. Explain the probability.

_____

_____

_____

22 A bowl contains 24 marbles. Of these 24 marbles, 3 are white, 5 are pink, 4 are red, 6 are yellow, 2 are purple, and 4 are orange. If you reach into the bowl without looking and choose one marble, what is the probability of choosing a white or pink marble?

_____

Example 4

Use the spinner to find the probability of spinning an even number. Write the probability as a fraction in simplest form. Explain the probability.

1. Count the number of sections labeled with an even number. Write this number in the numerator.

2. Count the total number of sections. Write this number in the denominator.

$P(\text{even number}) = \frac{4}{8}$ — The spinner has four even numbers.

The spinner has 8 numbers.

3. The ratio is $\frac{4}{8}$. The ratio can be simplified.

$$\frac{4 \div 4}{8 \div 4} = \frac{1}{2}$$

4. The probability of $\frac{1}{2}$ means that 1 out of every 2 spins should be an even number.

Write the ratio of width to length for each rectangle as a fraction in simplest form.

1 [rectangle, 6 cm height, 8 cm width] _____

2 [rectangle, 11 ft height, 3 ft width] _____

Write each ratio as a fraction in simplest form.

3 21 out of 168 were not wearing team colors _____

4 7 of the 15 fish in the tank were goldfish _____

5 21 laps in 3 days _____

Fill in the blanks and complete the ratio table.

6

Numerator	2	6	30	
Denominator	5	15		150

7

Numerator	24		6	2
Denominator	18	9	3	

Write each rate as a fraction. Find each unit rate.

8 60 miles in 2 hours _____

9 12 pounds in 3 weeks _____

Which product has the lowest unit cost? Round to the nearest cent.

10 8-oz bag of chocolate chips for $1.99, a 12-oz bag of chocolate chips for $2.49, or a 16-oz bag of chocolate chips for $2.99

_____

11 4 oranges for $1, 10 oranges for $2, or 24 oranges for $6

_____

12 4 DVDs for $59.99, 6 DVDs for $74.99, or 10 DVDs for $99.99

_____

Use the spinner to find each probability. Write the probability as a fraction in simplest form.

13 P(odd) _____

14 P(number greater than 4) _____

15 P(1 or 7) _____

Solve.

16 **READING** Toby was reading a book for his social science class. He read 144 of the book's 200 pages. Write the pages Toby has read to the total number of pages as a ratio in simplest form.

_____

17 Ms. Viera's class is taking a field trip. She needs to know how many vans they will need for the 32 students in her class. One van will hold six students. Complete the ratio table to solve the problem.

Vans	1	2	3	
Students	8	16		32

Correct the mistakes.

18 The Good Foods Market placed the sign to the right in the window. Is the sign correct?

Peanuts 8-oz bag that's less than $5.00 per pound

$2.99

_____

_____

19 When Javier and his sister Marissa were playing a game, Javier shuffled the deck of 52 game cards and asked Marissa to draw one from the deck without looking. He told her that she had a $\frac{2}{25}$ chance of drawing a red card. (There were 4 red cards in the deck of 52 game cards.) Is he correct?

_____

_____

STOP

Test Practice

Choose the best answer and fill in the corresponding circle on the sheet at right.

1 In Ruthie's bookshelf, there are 32 mysteries, 8 nonfiction titles, and 6 science fiction novels. What is the ratio of mysteries to nonfiction books?

A $\frac{16}{3}$, 16:3, or 16 to 3

B $\frac{4}{1}$, 4:1, or 4 to 1

C $\frac{1}{4}$, 1:4, or 1 to 4

D $\frac{16}{7}$, 16:7, or 16 to 7

2 Vik is driving to his grandmother's house. He makes the 275-mile trip in 5 hours. What is Vik's average speed?

A 1,375 miles   C 55 miles/hour

B 5 hours   D 65 miles/hour

3 A store sells an 8-pack of water for $4. What is the cost of one bottle of water?

A $0.50   C $2.00

B $1.00   D $4.00

4 Mr. Sato and his 26 students are going to an art museum. Admission and lunch for everyone will cost $364.50. What is the price per person?

A $13.50   C $14.50

B $14.02   D $15.00

5 Marta finished reading a novel in 8 days. The book was 384 pages. About how many pages did she read per day?

A 48 pages   C 96 pages

B 64 pages   D 112 pages

6 Samuel is running in a 26.2-mile marathon. If he completes the marathon in 4 hours, what rate did he average?

A 26.2 miles   C 5.15 miles/hour

B 4 hours   D 6.55 miles/hour

7 Which number completes the ratio table below?

Books	7	14	21	28
Shelf	15	30		60

A 45   C 21

B 40   D 50

8 A bag contains 5 blue, 6 white, and 3 red marbles. A marble is drawn without looking. What is the probability of drawing a white marble?

A 5 out of 14   C 6 out of 14

B 3 out of 14   D 6 out of 12

9 A number cube with six sides labeled 1 through 6 is rolled. What is the probability of landing on an even number?

A $\dfrac{1}{6}$

B $\dfrac{2}{6}$ or $\dfrac{1}{3}$

C $\dfrac{3}{6}$ or $\dfrac{1}{2}$

D $\dfrac{5}{6}$

---

10 Write a ratio that compares the number of faces to the number of hearts.

A 1 to 1

B 1 to 4

C 4 to 1

D 1 to 5

---

11 Which ratio could appear in the ratio table below?

Numerator	6		18	24
Denominator	5		15	20

A $\dfrac{7}{6}$

B $\dfrac{6}{10}$

C $\dfrac{12}{10}$

D $\dfrac{15}{15}$

12 Christopher tosses a coin. What is the probability that the coin will land on heads?

A 2 out of 2

C 2 out of 1

B 1 out of 2

D 1 out of 1

ANSWER SHEET

Directions: Fill in the circle of each correct answer.

1 Ⓐ Ⓑ Ⓒ Ⓓ
2 Ⓐ Ⓑ Ⓒ Ⓓ
3 Ⓐ Ⓑ Ⓒ Ⓓ
4 Ⓐ Ⓑ Ⓒ Ⓓ
5 Ⓐ Ⓑ Ⓒ Ⓓ
6 Ⓐ Ⓑ Ⓒ Ⓓ
7 Ⓐ Ⓑ Ⓒ Ⓓ
8 Ⓐ Ⓑ Ⓒ Ⓓ
9 Ⓐ Ⓑ Ⓒ Ⓓ
10 Ⓐ Ⓑ Ⓒ Ⓓ
11 Ⓐ Ⓑ Ⓒ Ⓓ
12 Ⓐ Ⓑ Ⓒ Ⓓ

Success Strategy

Easier questions usually come before harder ones. For the more difficult questions, try to break the information down into smaller pieces. Make sure the answer is reasonable and matches the question asked.

STOP

Index

Copyright © Glencoe/McGraw-Hill, a division of The McGraw-Hill Companies, Inc.